· Environmental Auditing ·

Environmental Auditing

Neil Humphrey
BSc (Hons), MSc
Director of Environmental Auditors Ltd and ContamiCheck Ltd

and

Mark Hadley
BSc (Hons)
Managing Director of Environmental Auditors Ltd

Foreword by
Martin Baxter
Institute of Environmental Management & Assessment

Palladian Law Publishing Ltd

© Neil Humphrey and Mark Hadley
2000

Published by
Palladian Law Publishing Ltd
Beach Road
Bembridge
Isle of Wight
PO35 5NQ

www.palladianlaw.com

ISBN 1 902558 26 X

Typeset by Heath Lodge Publishing Services
Printed in Great Britain by The Cromwell Press Ltd

· Contents ·

About the Authors   vii
Foreword   xi
Preface   xii
Acknowledgements   xiii
Table of Statutes   xiv
Table of Statutory Instruments   xv
Abbreviations   xvi

1   Introduction   1
    Definition of "environmental auditing"   1
    Types of auditing   4

2   UK and EU Environmental Legislation   15
    Introduction   15
    Environmental legislation – the European context   16
    Principal sources of UK legislation   21
    UK environmental and regulatory authorities   48

3   The Audit Process   58
    Introduction   58
    Pre-audit activities   58
    On-site audit activities   61
    Post-audit activities   67

4   Collecting Background Information   69
    Determining a site's past and present uses   69
    Historical maps   69
    Planning records   70
    New information sources – environmental data providers   81
    Other environmental sources   94
    Collecting information about a site   97

5   Audit Methodologies and Working Papers   104
    Introduction   104
    Audit protocols   104
    Questionnaire protocols   108
    Working papers   111

6   Identifying Environmental Effects   116
    Operations   116

	Operational impact areas	118
	Process diagrams	124
7	**Reporting and the Audit Process**	**127**
	Introduction	127
	Oral reports	127
	Formal written reports	128
	Action list	131
	Criteria for reporting	131
8	**Contaminated Land, Groundwater and Pollutant Pathways**	**132**
	Introduction	132
	Contaminative risks	133
	Sources of contamination	140
	Contaminated groundwater and surface water	142
9	**Site Remediation**	**146**
	Introduction	146
	Commonly used methods for remediation – groundwater	147
	Commonly used methods of remediation – soil	149
10	**Developments in Environmental Auditing**	**151**
	Introduction	151
	EN ISO 14010:1996	151
	EN ISO 14011:1996	153
	EN ISO 14012:1996	154
	Glossary of Terms	157
	Bibliography	166
	Useful Contact Numbers and Information	169
	Appendix 1: Pre-acquisition Due Diligence Inquiry	178
	Appendix 2: IPC and Air Pollution Guidance	192
	Appendix 3: Proposed Phase-in Dates	199
	Appendix 4: DoE List of Contaminative Uses	201
	Appendix 5: Pre-survey Questionnaire	206
	Appendix 6: Contaminated Land Regime	210
	Appendix 7: Guidance on Assessment and Redevelopment of Contaminated Land	215
	Index	221

· About the Authors ·

Neil Humphrey

Neil Humphrey was born in Brighton in 1971. Having graduated from Kings College, London with a BSc (Eng) honours degree in engineering and management he started working in marketing, realigning himself later in the environmental field. He subsequently graduated with an MSc in Marine Resource Development and Protection from Heriott Watt University, living both in Orkney and Edinburgh during the course of his studies. He joined Environmental Auditors Ltd (EAL) in 1995, specialising in undertaking marine environmental impact assessments and soon became interested in the field of environmental auditing and contaminated land. In 1996, Neil presented a number of lectures at the University of Brighton focusing on the practical aspects of undertaking environmental assessments and in 1997 managed a joint venture with EAL and the University in developing an Environmental Auditors Registrations Association (EARA)-recognised foundation course in environmental auditing. Neil wrote the majority of the text for the course, which is currently run from the University on an annual basis. In 1997, Neil was promoted to Operations Manager at EAL, and in 1998 was appointed to EAL's Board of Directors. Neil is an EARA-registered environmental auditor and has extensive experience in undertaking due diligence environmental audits, contaminated land investigations and site remediations. Neil is also a director of ContamiCheck Ltd, a company offering low cost, desk-top assessments using archived database information (including all the Environment Agency data for the UK), and has assisted on a number of expert witness cases for the Environmental Law Foundation.

Mark Hadley

Mark Hadley was born in Worcestershire in 1958. He graduated from Imperial College of Science and Technology (London University) with a BSc Hons degree in zoology and commenced working as a scientific officer for the scientific civil service. After five years, he left to found Bioscan (UK) Ltd, one of the first ecological consultancies in the United

Kingdom. Initial experience was gained with environmental impact assessment work for a number of major oil companies, followed in 1986 by his departure to join Tecnitas SA, the technical consultancy arm of Bureau Veritas SA. He was elevated to Managing Director of Tecnitas (UK) Ltd, and built a substantial consultancy operation based on environmental impact assessment work throughout the world. In 1989 he joined a firm of process engineering consultants, Trident Consultants Ltd, as their environmental manager, where he honed his skills on environmental impact assessment for the petrochemical and chemical industries. During this period, Mark also developed his skills in the developing environmental auditing industry, formulating techniques for waste management audits, as well as acting on a seconded basis as environmental advisor to Gulf Oil (GB) Ltd.

In 1991, Mark joined the growing team at Central and Provincial Management Ltd, a specialised property consultancy managing over 2,000 industrial properties based in Mayfair, London. He formed a division of the company, Environmental Auditors Ltd, applying specialised skills in auditing, risk assessment and liability appraisal for major blue-chip industrial clients. In 1993, he successfully concluded a management buy-out of the embryonic business, relocated it to Sussex, and started building a business based on auditing and related consultancy services.

He is currently managing director of EAL, as well as a key player in associated businesses including an environmental data business, ContamiCheck Ltd, a company specialised in providing environmental impairment liability insurance (ERIS). He was co-founder of the original EARA and is an accredited Principal Environmental Auditor as well as a course examiner for EARA, and a Certified Environmental Inspector (US). Mark has variously served as the chairman of the publications committee and council member for the Pipeline Industries Guild, the Institute of Environmental Assessment and is a member of the Institute of Petroleum and the Association of Insurance Risk Managers. He acts as an expert witness on contaminated land and pollution issues for the Environmental Law Foundation and a number of legal organisations and is an external examiner for the University of Brighton.

Environmental Auditors Ltd

EAL is an independent consultancy which has, for a number of years, been providing a wide range of environmental services to the property, insurance and banking industries, P & I clubs, multinational corporations and to the manufacturing and service sectors. EAL provides a flexible, responsive and effective service, and has established a reputation for the quality of its work and for its objective and impartial approach. The company operates within the United Kingdom but also has considerable experience internationally, with work being completed in a number of countries including Greece, Brazil, France, Spain, Belgium, South Yemen Republic, India, Korea, Malaysia, Indonesia and Mexico.

EAL is a corporate member of the Institute of Environmental Assessment and has close links with a wide range of other bodies including the Institute of Petroleum, the Institute of Environmental Sciences, the Association of Insurance and Risk Assessment Managers, the Institute of Water and Environmental Management, the Institute of Waste Management, the Institute of Gas Engineering and the Royal Society of Chemistry.

· Foreword ·

Environmental auditing continues to develop as an essetnial and integral component of the environmental management profession. As organisations become more aware of the environmental implications of their activities, products and services, greater attention is being paid to the way they manage and improve their environmental performance. As a mangagement tool, environmental auditing has developed to provide information on how well, or how badly, organisations are meeting their environmental policies. It is also an essential component of the checks that are now an everyday part of land, property and business acquisitions and divestments.

Environmental legislation and regulation are becoming ever more stringent and complex, and the penalities for failure ever more serious. The need for skilled, competent and professional auditors is paramount. The more recent pressure for regulatory recognition of formal environmental management systems, such as the EC Eco-Management and Audit Scheme (EMAS) and the international environmental management system standard ISO 14001, means that closer attention will be given to the competence of auditors and the results of the audit process. Organisations can only expect to receive any form of regulatory relief or dividend when they can demonstrate to the regulators that environmental auditing can act as an effective and reliable self policing system.

This timely book provides an insight into the audit process and the skills and attributes that comprise a competent auditor. It is a welcome publication that will provide a valuable contribution to the developing literature in this important subject.

Martin Baxter
Professional Standards and Development Manager
Institute of Environmental Management and Assessment

The Institute of Environmental Management and Assessment was formed from a merger between the Environmental Auditors Registration Association, the Institute of Environmental Management and the Institute of Environmental Assessment.

· Preface ·

This book has in the main been drawn from the text and workshop exercises of Environmental Auditors Ltd's EARA accredited Environmental Auditors Foundation Course, together with other materials collected over the years as a practical environmental company which is foremost in the field of environmental auditing.

The course has been presented in association with the University of Brighton for the last three years and has aimed to provide its delegates with both the essential skills and knowledge to effectively undertake "Phase 1" environmental audits.

The aim of this book is to provide an authoritative source of reference for both professionals and students who through the course of their work or studies have an interest in the concept of environmental auditing. This book is therefore an introduction to the field of environmental auditing, to be used as a guide and aide-mémoire within the auditing process. The authors have between them extensive experience of the auditing process and as such it is hoped that this book provides a useful source of reference. The principal benefit of this book is that its content is recognised by the EARA within the remit of a recognised foundation course, and as such all the essential information in enabling an individual to undertake an audit should be contained within these pages.

Because of the extensive range of literature and reference material currently available, the impetus of this book is not necessarily "what is an environmental audit?", but rather how an is audit undertaken; the key questions that should be asked and what to look for in an audit; and the key issues which should be established within the remit of an audit. It is hoped that this book fulfils these objectives.

For the sake of clarity, note that only "he", "his" and "him" are used in the text; such references also refer to the feminine pronoun.

Neil Humphrey
Mark Hadley
December 1999

· Acknowledgements ·

Initial thanks has to go to Nick Lightbody for introducing Mark and myself to Palladian Law Publishing. Specific thanks must also go to Andrew Prideaux at Palladian, who has shown amazing amounts of understanding and patience throughout the entire project, even when we were repeatedly sidelined and had to delay the writing of this book. Thanks must also go to the many people who have contributed to this book, both directly and indirectly. Although we attempt to list them all, we have undoubtedly omitted one or two people and for that we apologise and say "Thanks". On that basis, thanks to Denise Hill and Jean Lyle at the University of Brighton, and not forgetting Mark Pendry and Guy Mercer (previous employees of EAL) for all their hard work in the original development of the EARA course, from which we doubt this book would have been possible. Thanks to everyone at EARA, specifically Ruth Bacon and Richard Hook for their assistance during the approval process of the EARA course (although I may not have said "Thank you" at the time). Special thanks also to Ruth for her later help and input during the writing of this book. Thanks to everyone at EAL for their help and support whilst writing this book, specifically Matthew Greaves, James Hanson, Simon Turner, Shelley Moyes, Amy Hines, and Chris Taylor (specifically for his input into Chapter 4).

Publishers' Note. The publishers and the authors are grateful to the following for permission to reproduce material: Information for Industry Ltd for the list of contacts published in the *Environmental Compliance Manual*, Ordnance Survey for the maps and the Controller of Her Majesty's Stationery Office for Crown copyright material.

Neil Humphrey

· Table of Statutes ·

Alkali Act 1863   15

Control of Pollution Act 1974
  15, 30, 34
  s 30A(d)   158
Control of Pollution
  (Amendment) Act 1989   32

Environment Act 1995   36
  s 80   41
  ss 82-84   41
  s 92   41
  Part I   41, 55
  Part II   41
  Part IIA   41, 42
  Part III   41
  Part V   13
Environmental Protection Act
  1990
  s 1   23
  s 27   24
  s 29(3)   30, 31
  s 33(1)(c)   30, 31
  s 34   12, 31, 158
  s 75(2)(4)   33
  s 78   184
  s 78A   42
  s 78 B(1)   43
  s 78 F   45
  s 78 J(3)   46
  s 78 K   46
  s 78 X(4)   46
  s 79(1)   164

Part I   21, 22,27
Part II   22, 31
Part III   22, 33
Parts IV-IX   22
European Communities Act
  1972   17

Health & Safety at Work Act
  1974   39

Planning (Hazardous Substances)
  Act 1990   37, 38
Pollution Prevention & Control
  Act 1999   22, 25, 30
Public Health Act 1993   15

Sewerage (Scotland) Act 1968
  s 59(1)   164
Single European Act 1986
  s 130   17

Town & Country Planning Act
  1990   70

Water Industry Act 1991
  s 118   35
  s 141   164
Water Act 1989   34

Water Resources Act 1991   34
  s 85(1)   35
  s 104(1)   158

Table of Statutory Instruments

Anti-Pollution Works Regulations 1999   36

Contaminated Land (England) Regulations 1999   44, 210
Control of Industrial Major Accidents Hazardous Substances
  Regulations 1984   37, 101, 120, 122
Control of Major Accident Hazard Regulations 1999   38
Control of Substances Hazardous to Health Regulations
  1988   38, 39, 120
Controlled Waste (Registration of Carriers and Seizure of Vehicles)
  Regulations 1991   32
Controlled Waste Regulations 1992   33

Environmental Protection (Duty of Care) Regulations 1991   32
Environmental Protection (Prescribed Processes and Substances)
  Regulations 1991   23, 30

Groundwater Regulations 1998   35

Notification of Installations Handling Hazardous Substances
  Regulations 1982   37, 101

Planning (Hazardous Substances) Regulations 1992   37
Pollution Prevention & Control (England) Regulations 1999   22, 25
Pollution Prevention & Control (Scotland) Regulations 1999   22, 25
Producer Responsibility Obligations (Packaging Waste) Regulations
  1997   119

Special Waste Regulations 1996   33

Town & Country Planning (Hazardous Substances) (Scotland)
  Regulations 1993   37

· Abbreviations ·

ASTM	American Society for Testing and Materials
BAT	best available technique
BATNEEC	best available technique not entailing excessive cost
BPEO	best practical environmental option
CBI	Confederation of British Industry
DETR	Department of the Environment, Transport and the Regions
DoE	Department of the Environment (now the DETR – see above)
DTI	Department of Trade and Industry
EAL	Environmental Auditors Ltd
EARA	Environmental Auditors Registration Association
EDA	Environmental Data Association
EMS	environmental management system
EPA	Environmental Protection Act
EU	European Union
HMIP	Her Majesty's Inspectorate of Pollution
HSE	Health & Safety Executive
ICC	International Chamber of Commerce
IEO	Industry and Environment Office
ICRCL	Interdepartmental Committee on the Redevelopment of Contaminated Land
IPC	integrated pollution control
IPPC	integrated pollution prevention and control
IRF	information request form
LAAPC	local authority air pollution control
MAFF	Ministry of Agriculture, Fisheries and Food
NRA	National Rivers Authority
OECD	Organisation for Economic Co-operation and Development
PSQ	pre-survey questionnaire
UNEP	United Nations Environmental Programme

Chapter 1

· **Introduction** ·

The concept of environmental auditing developed in the mid- to late 1970s in the United States. The technique was initially pioneered by the country's largest multinational corporations such as Allied Signal and General Motors as a response to an increasingly complex environmental regulatory framework. The main function of auditing at this time was to determine the extent of a company's compliance with various state and federal environmental legislation. A particular fillip was given in the 1980s for ensuring such compliance when an Environmental Crime Unit was established by the US Department of Justice. The purpose of this unit was to convict businessmen who either failed to comply with, or who failed to ensure that their subordinates complied with, environmental legislation.

From its origins in the United States, environmental auditing has spread throughout the world. Primarily, this spread has been owing to the commitment of the American multinational companies to the auditing process. As the environmental policies of such organisations apply to all their operations, environmental auditing was extended to the activities of their overseas subsidiaries. Hence the practice of auditing was introduced into many different countries. With its introduction to the United Kingdom the auditing concept was rapidly adopted by a number of the country's largest companies, including ICI and Shell.

Over time the concept of auditing has evolved considerably, and the methodologies employed in auditing have developed likewise. As the discipline has developed a number of types of audit have evolved.

1.1 Definition of "environmental auditing"

In preparing a book on environmental auditing, it is perhaps wise to begin with a definition.

The term "environmental audit" has been in use since the 1960s, when the ITT Corporation referred to "environmental compliance auditing". Later, in 1975, the Allied Chemical Corporation was hauled before the US courts following environmental problems at its Hopewell, Virginia plant and told to obtain "an environmental audit program". That programme, and the many that have subsequently followed, learned much from systems developed for reviewing health and safety management in the industrial and commercial sector. An environmental audit, however, is different: while the name itself sounds business-like and technical, the term has been actively employed to cover a whole range of environmental assessment and review procedures.

To audit, as defined by the *Oxford English Dictionary*, is to "make an official systematic examination (of accounts), so as to ascertain their accuracy"; there is no convenient definition of what constitutes the environmental equivalent.

It is possible to take on board some of the North American definitions which, while rather long-winded, do provide some guidance. In 1981, the US Controller General suggested that an environmental audit constituted: "Work done not only by accountants and auditors in examining financial statements, but also work done in reviewing compliance with applicable laws and regulations, economy and efficiency of operations, and effectiveness in achieving program results". This quickly evolved into a formal definition that supported the US Environmental Protection Agency Environmental Auditing Policy Statement issued in July 1986, which read: "Environmental Auditing is a systematic, documented, periodic and objective review by regulated entities of facility operations and practices related to meeting environmental requirements", those entities being private companies or public bodies subject to environmental regulations (UNEP/IEO, 1990). In other countries (*e.g.* Canada), private companies such as Noranda Inc stated:

> "An Environmental Audit Program is the key element of any Environmental Management System, which requires external auditing as a systematic and objective method of verifying that ... environment, health, industrial hygiene, safety and emergency preparedness standards, regulations, procedures and corporate guide-lines are being followed."

(UNEP/IEO, 1990.) Other global corporations such as 3M tried to simplify the definition by developing several definitions, each for individual needs; thus, "operational auditing" was "examining everyday operations and how they match up with legislation, company standards and good practice". When 3M turned its attention to acquisition audits (*i.e.* audits carried out prior to the purchase of a property or company), it failed to provide a definition. This really is the essence: how does one define and then carry out audits relevant to property transactions and corporate activities where environmental liabilities need to be considered?

As with much environmental legislation, North America was developing systems and strategies for evaluating such considerations long before Europe. It would be fair to say that the term "environmental audit" travelled across the Atlantic in a piecemeal fashion, often used by North American companies that were adapting their environmental review and management systems for use in Europe, and in particular the United Kingdom.

Auditing is an activity of verification, the comparison of outcomes against expectations. Where such standards or expectations do not exist, the environmental audit becomes an assessment.

Because the use of the term has become so widespread, the International Chamber of Commerce (ICC) laid down a position paper on environmental auditing which was published in March 1990 (UNEP, 1990). The definition that was arrived at is now widely accepted throughout both the United Kingdom and the rest of Europe, and states that environmental auditing is:

> "A management tool comprising a systematic, documented, periodic and objective evaluation of how well environmental organisation, management and equipment are performing with the aim of safeguarding the environment by:
> (i) facilitating management control of environmental practices; and
> (ii) assessing compliance with company policies, which would include meeting regulatory requirements."

This definition outlines a number of distinctive features of environmental auditing:

(1) *Audits should be a systematic and comprehensive.* To achieve this audit protocols have to be developed. These protocols outline all the procedures and actions to be undertaken during an audit and guide the auditor through the audit.
(2) *Audits should be fully documented and, where possible, substantiated with physical evidence.* Such documentation is essential as it provides a means of supporting any conclusions and findings made during the audit. Documentation of an audit can be achieved principally by the preparation of working papers.

(3) *Audits should be periodic.* Audits should not generally be a "one-off" procedure, but rather should be seen as a dynamic and frequently repeated process.

(4) *Audits should be objective.* It is essential that an audit provides a true indication of the situation at a site or within a company. It is essential, therefore, that the auditor is impartial, objective and is independent of the site or company being audited.

This is an excellent definition for operational reviews of companies or individual sites, but is too restrictive when applied to the wide range of audits being performed for the purposes of property transactions. It is for that reason that Dr Tim Coles of the Institute of Environmental Assessment drafted an elegantly simple definition in the Institute's *Preparatory Review of Auditing*, issued in May 1992. This states: "Environmental Auditing is the monitoring of a company's performance against previously agreed policy or statutory standards".

It is true that some organisations have been resilient to the use of the term "audit", with British Airways using "Review", Allied-Signal using "Environmental Surveillance Programme" and the EU using "Eco-Audit". However, a consensus has been reached with respect to the term "environmental audit" which has quickly established itself in the corporate culture of, for example, British Coal, Dow Chemical Co Ltd, Rhône-Poulenc, Ciba-Geigy Ltd, ITT Corporation, Shell International Petroleum, Norsk Hydro, BP International Ltd, Coopers & Lybrand Deloitte, Commercial Union Assurance Co, Proctor & Gamble, Body Shop and 3M.

1.2 Types of auditing

While many different organisations tried to produce a viable working definition of exactly what an environmental audit was, both the public and private sectors of industry, as well as local government, went ahead with their own auditing programmes. They each designed a methodology for carrying out an audit dependent on their own needs and objectives which has resulted in a plethora of audit types.

It is necessary to have a basic understanding of the different types of auditing that are carried out. A simple example of this is the design audits developed by BREEAM (Building Research Establishment Environmental Assessment Method) specifically to assist the construction industry in its task of creating "environmentally friendlier" buildings. Great account of this has been taken in the commercial property sector where new offices, superstores and residential buildings are being held up for comparison against the standard.

In carrying out an environmental auditing programme, three fundamental considerations need to be borne in mind:

- Structural considerations
- Resource considerations
- Issues to be audited.

Structural considerations cover the physical scope of the audit; generally speaking these environmental audits usually examine either specific sites or whole organisations. The organisations may be separate operating companies or statutory authorities, such as a local authority. When these organisations are extremely large (*e.g.* multinational companies such as the Royal Dutch Shell Group, British Petroleum International or IBM Ltd), it would be far too onerous a task to audit the whole organisation quickly, given that it may have many hundreds of subsidiaries operating in dozens of countries, making a single meaningful audit impractical. Thus in these very large organisations the emphasis in auditing has shifted to the functional, with the specific audit programme designed to report across a range of companies on different functional aspects such as operational health and safety, purchasing policy, transport policy or communications.

An audit can be carried out with internal resources, such as the company's own employees; these are often selected from different parts of the organisation to those being audited. This is very much the approach pursued by multinationals which often take teams of experts from one part of the body corporate, and set these to task auditing the different subsidiaries in the group. Alternatively, it is possible to source the expertise outside the organisation, from external consultants or from consultancy companies. There are advantages and disadvantages to each approach, and it is for this reason that many organisations have integrated both types of approach by using their own employees but under the direction and control of an external lead auditor. The choices available are dictated by available resources and the nature of the organisation involved.

Much has been published in both Europe and North America on the classification of audit types. They share one thing in common, they are (or should be) demand led. Before embarking on an audit programme it is necessary to choose the most appropriate type of audit, and this will be dependent upon the objectives initially prescribed.

Basically, there are three mainstream types of "corporate" audit: compliance audits, single issue audits and liability audits. The relationship between these is shown in simplified form in Figure 1.1.

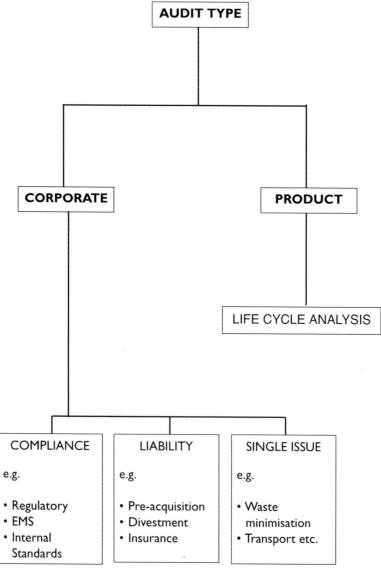

Figure 1.1: Relationship between the different auditing types

During the evolution of auditing technique, methods and protocols have been established which satisfy the broad requirements of achieving compliance with statutory requirements, identifying the environmental liabilities or quantifying costs associated with waste generation and energy use. Figure 1.1 attempts to place these in the three broad categories, and while some types of audit fall clearly into one category,. the majority of audits fall between two categories and might therfore be referred to as "hybrids". There is nothing wrong with this concept – indeed the needs of different organisations with varying objectives suggests that many of the audit programmes that have developed have a set of objectives which range between compliance on the one hand and avoidance of liability on the other. Furthermore, these audits, depending on their scope, often cover many of the subjects suggested as suitable for single issue audits.

Compliance audits

Compliance auditing is a technique for assessing an organisation's ability to comply with local or national environmental regulations (or, in North America, with state and Federal regulations), or alternatively with an organisation's own internal environmental policy. In the latter case, internal company policies usually require that all existing legislative requirements are fully complied with anyway. The structure of a compliance audit varies considerably between different types of organisation: even companies in the same industrial sector can have widely differing compliance audit systems, and it is perhaps because of concerns about the levels of control established by such systems that more formal arrangements have been established.

While at present environmental auditing has not become mandatory in the United Kingdom or elsewhere in the world, protocols have been elucidated to formalise the structure and content of such assessments. On 9 July 1986 the US Environmental Protection Agency issued its *Environmental Auditing Policy Statement*, which sought to "encourage the use of environmental auditing by regulated entities to help achieve and maintain compliance with environmental laws and regulations, as well as to help identify and correct unregulated environmental hazards".

Outside the United States, roughly the same conclusions were reached by a number of different organisations. In March 1989, the ICC published its *Position Paper on Environmental Auditing*, which described the basic elements of environmental auditing and sought to suggest how the technique could be best employed in improving the effectiveness of health, safety and environmental programmes. This Position Paper was made available from the ICC's National Committees situated in 59 countries and established a marker for many of the publications that were to follow later.

In the United Kingdom, the Confederation of British Industry (CBI) produced a small booklet in June 1990 on environmental auditing in its "Guide-lines for Business" series. While these publications were useful contributions to raising the level of environmental awareness, they were short on technical detail.

One of the first detailed protocols to be promoted was the *Guidance on Safety, Occupational Health and Environmental Protection Auditing*, which was issued by the Chemical Industries Association in February 1991. While this was a sectorial approach to the subject, it managed to produce at least 10 pages of good, practical guidance on aspects of environmental compliance. The guideline provided a framework for both legislative and policy compliance, but was also a hybrid audit technique being expansive enough to cover health and safety issues as well.

Individual companies such as Dow Chemical Company followed this guidance. They made a public commitment to the Chemical Industries Association's "Responsible Care" Programme which covers all aspects of chemical manufacture, transport, use and disposal, thus complying with an industry-wide policy objective, their own corporate policy objectives and all current standards and regulations. Most compliance auditing is a composite of policy and legislative objectives, which may or may not extend into areas of occupational health and safety.

Advantages

There are inherent advantages in consolidating these various issues:

(1) It saves time while on site, with the audit team only having to make a single visit, although the time for this is extended, mobilisation and demobilisation procedures are only necessary once.
(2) It reduces duplication when several teams all assess common aspects such as management structure, policy, communication, maintenance procedures and incident reporting.
(3) It allows auditors with a wide variety of backgrounds and training to consider specific problems as they might affect environmental, health and safety aspects.

It is not uncommon to find that a modification to plant or process will reduce the environmental impact of that activity, but has a negative safety aspect, or vice versa. These can be discussed, evaluated and mitigated against within an interdisciplinary audit team.

Disadvantages

The disadvantages with the multi-functional approach to auditing include:

(1) Complex logistics: the larger the size of the audit team, or the involvement by more than one organisation within the team, may lead to difficulties in getting all the appropriate expertise to the site within the dates specified.
(2) There may be "management resistance", since a large multi-disciplinary team will be very demanding of management time, and may involve several key personnel at one time, and thus interfere with the smooth-running of the company being audited.
(3) Extended auditing programmes may take several days or weeks; this is particularly true of complex manufacturing or processing plants where the complexity of the issues may occupy a large team for two or three weeks on site. This can be very wearing on the individual team members, particularly where they are not directly concerned with, or experienced in, certain areas of expertise, and "auditor fatigue" sets in.

It is for these reasons that many organisations have preferred to keep environmental auditing separate from existing health and safety audit programmes.

Single issue audits

Single issue audits require the minute examination of one particular aspect of a company's or factory's operations. They are often triggered by a sudden management realisation that bottom-line cost savings can be made by increasing energy efficiency or reducing waste generation. It is not uncommon to find that, once a company has been through a general programme of compliance auditing, specific "single issues" are made the subject of more detailed scrutiny.

Liability audits

This is a large category encompassing a wide variety of audit protocols. They all, however, share a common objective: the desire of the commissioning party to avoid environmental liability. There are three main types of liability audit which are characterised by the type of commissioning party:

- Pre-acquisition audit
- Divestment audit
- Insurance audit.

These three main types have all been given at various times different names, but the purpose for which they are commissioned remains the same. There are also some hybrid audits which have strong liability connotations, and these are also discussed below.

Pre-acquisition audits

These are also referred to as "purchaser", "pre-merger" and "due diligence" audits. They are usually commissioned by companies in advance of purchasing other companies, specific properties or occasionally land that has a known history of industrial use, or prior to a merger with another company where property or land holdings are involved.

Increasingly nowadays, lending institutions providing the capital necessary for such a purchase or merger will demand that a pre-acquisition audit is carried out by the borrower. The audit will provide a measure of comfort to the lender that the loan is secure, and that any potential financial consequences arising out of the past or current use of the property will not affect its security. A natural extension to this policy of scrutiny by banks and other lending institutions has been to carry out a similar audit prior to accepting the surrender of industrial or commercial properties where latent environmental liability is suspected. The objective is simple: to determine whether there are any environmental liabilities associated with the acquisition that would result in the property or company being worth less in the future, or at worst leading to closure of the company or factory because of regulatory intervention or future pollution emanating from the site.

It is normal during any property transaction for the purchasing party to request specific answers to a set of due diligence inquiries made by its solicitors. These have been extended in recent years to cover environmental issues, with the pre-acquisition audit or review acting as a useful tool for collating this information. Typical pre-acquisition due diligence inquiries have been developed by The Law Offices of N F Lightbody, and are presented in Appendix 1.

Divestment audit

A divestment audit, which is effectively the opposite of the pre-acquisition audit, has similarly been referred to as "vendor", "disposal" or "de-merger" audit. Once again the objective is to ascertain the specific environmental liabilities with respect to property and land at a specific site or sites, or as part of the sale and disposal of a subsidiary company. However, the objectives differ from those of the pre-acquisition audit. In this case, the vendor may have several reasons for commissioning such an audit. Perhaps one of the most frequent reasons given is that it will establish a baseline of information against which the advantages and disadvantages of any transaction can be gauged. The vendor instructs the collection and collation of all the environmental information that will be necessary to satisfy the questions asked by potential purchasers. It remains incumbent on the potential purchasers to check that the information is both adequate for their purposes and accurate. This will save time during the transactions, and demonstrate to potential purchasers that the vendor is aware of potential environmental liabilities and is prepared to be quite open and candid in its future negotiations.

Alternatively, it is a means by which the vendor can "cap" its liabilities by providing all pertinent information on the condition of the facility or company being sold. This ensures that the purchaser is cognisant of the liabilities attached and will in effect, protect the vendor from potential claims in the future.

Quite often, vendor audits are commissioned prior to site disposal to provide a reference point against which the questions raised in the purchaser's pre-acquisition audit can be answered and, if necessary, challenged.

Insurance audit

The reputation and profitability of a company can be seriously affected, in both the long and the short term, by third party liability claims. These claims may result from actions taken by staff employed at, customers of and neighbours adjacent to a specific site and are most usually a result of pollution liability claims.

To control, and hopefully limit, the extent of the insurer's exposure it is necessary to understand and to measure the levels of risk posed by individual companies or sites. Insurance underwriters are increasingly requiring solid facts concerning the probabilities of sudden and accidental pollution arising together with some assessment of the potential magnitude of such incidents, should they arise.

Those companies and organisations which can demonstrate clearly that they have adequate systems for controlling risk, thus limiting liabilities, are rewarded with lower premiums or more extensive policy cover. In many cases, insurance will only be available for companies which can demonstrate that detailed risk assessment has been undertaken.

Systems for managing and controlling risks include a range of audit techniques specifically tailored to the insurance industry. For environmental impairment liability cover, an environmental audit can satisfy most of the basic information requirements. This it does by demonstrating the use of technologies and monitoring systems designed to avoid risks, control those risks when they occur and limit the potential scale of any pollution incidents that ultimately might occur.

Hybrids

Audits are often required to satisfy a number of different requirements, and liability audits have often been combined with compliance audits and single-issue audits.

In the United States and increasingly in the United Kingdom, environmental legislation has placed a requirement on the producers of waste materials (especially toxic wastes) to closely control storage, removal, transport and final disposal. In the United Kingdom, section 34 of the Environmental Protection Act (EPA) 1990 places a duty of care on anyone who imports, carries, keeps, treats or disposes of controlled wastes, and this applies to any properties used for the above purposes. Specific regulations came into effect in April 1992 which detail how the wastes should be controlled; these suggest it is advisable to carry out inspections of the final disposal sites of any such wastes.

An auditing system can be used to follow and control the movement and disposal of such wastes and ensure compliance with the regulations. This type of audit is a hybrid, providing assurance across a broad range of environmental risks while focusing upon a single issue, namely waste management.

Another, similar hybrid audit can be carried out on suppliers. It is the stated policy of the "do-it-yourself" retail group, B & Q plc to require all their suppliers to provide documentary evidence that the products they supply to the retail chain are produced in the most environmentally benign way. In attempting to meet this objective, suppliers are required to have an audit undertaken of their production operations and supply documentary evidence to B & Q plc. Suppliers failing to provide the necessary documentation to the agreed standard are considered at a disadvantage to other suppliers of the same products who are in compliance with the policy. Again, this is a hybrid audit looking at the environmental liabilities and consequences of product manufacture against a single policy requirement.

Product audit – life cycle assessment

The veritable explosion of interest in environmental concerns that occurred during the 1980s in the United Kingdom, and much earlier in North America, has focused on the basic impacts of certain industries and products on the environment. This has resulted in the so-called "green" audit, or life cycle analysis. In this context, the audit looks at the complete life history of a specific product, or range of closely related products. It charts the sourcing of the raw materials needed for manufacture, transport of these to the place of manufacture, manufacturing processes, use of the product and then final disposal and waste management. A thorough life cycle analysis requires minute examination of all the activities concerning a specific product, and is usually commissioned by a company or an organisation that wishes to demonstrate, usually publicly, that its activities are as environmentally sound as is possible within the given limits of current technology.

Life cycle assessment is based on a consideration of all the environmental effects of a product or system "from the cradle to the grave". The assessment therefore encompasses all the operations associated with the subject of the audit. In addition to consumer pressure, producers of goods are now responsible for the ultimate disposal of their goods under the EU Directive on Packaging and Packaging Waste (94/62/EEC), implemented by Part V of the Environment Act 1995. Consequently, for companies to manage both these commercial and legal pressures, there is a need for them to implement a tool with which they can assess the impacts of their products throughout the product's life cycle. One of the principle factors of life cycle assessment is the need to consider the environmental impacts of companies' operations in order to identify significant environmental effects and define priorities for performance improvement.

Upon goal definition and scoping of the life cycle assessment, the next step is the identification and quantification of all the inputs into and all the outputs from the product or system under study. The inputs and outputs need to be established wherever possible and traced to either an input to the natural environment or an output to the natural environment. These inputs and outputs are known as "elemental flows". As far as possible, each input or output substance should be separately identified and quantified.

As would be expected, problems may be encountered where a process produces more than one useful product and hence the allocation of inputs and outputs may be hampered. In addition, it may become increasingly difficult to undertake this assessment for the operations of a company's suppliers and customers.

As previously detailed, one of the principal objectives of this assessment is to identify the effects on the environment associated with a process, operation or system. This assessment can generally be divided into three stages:

- Classification
- Characterisation
- Valuation.

The classification stage involves associating each input or output with one or more environmental effects. Characterisation involves the weighting of different inputs and outputs in relation to their relative contributions to each environmental effect with which they are associated. Finally, valuation involves assessing the relative importance of the different environmental effects.

Chapter 2
UK and EU Environmental Legislation

2.1 Introduction

Before examining how to undertake an environmental audit, it is in the first instance important to gain an understanding of the present legislative environment, specifically when applied to environmental protection. Note that it is beyond the remit of this book to provide an authoritative interpretative account of the present UK environmental legislative system and consequently detailed interpretation of UK and EU law is not an objective of this chapter. Where interpretation has been provided, this interpretation is limited to the authors' own experience and is based on what may be considered as reasonable. On this basis, the reader is directed to the extensive literature currently available on interpretation of UK and EU law, taking full account of any court precedents set.

When the United Kingdom joined the European Community in 1973, it already had a significant proportion of legislation directed towards the environment. The earliest environmental legislation was implemented as far back as 1863 with the Alkali Act, which was aimed at controlling atmospheric emissions primarily from the caustic-soda industry. After this, environmental legislation developed in a very haphazard way, being contained in a number of different areas of legislation such as the Town and Country Planning Acts, Clean Air Act and Public Health Act (Side, 1993).

The Control of Pollution Act 1974, was the first major piece of legislation specifically aimed at environmental pollution control. The Act pulled together all aspects of legislation relating to environmental pollution. However, although it was very detailed, the Act experienced problems in terms of its implementation and it was soon realised that there was a need for sharper and more extensive controls over certain pollutants.

In tandem with the development of UK environmental legislation, the EU has developed its own coherent body of legislation and controls relating to the protection of the environment, and most UK legislation is based on the implementation of these EU Directives and Regulations. It is therefore important to appreciate how EU environmental policy has developed and in turn how this has been translated into the UK legislative framework.

2.2 **Environmental legislation – the European context**

The Treaty of Rome in 1957 established the European Economic Community and provided the framework for common social and economic policies between the Member States. It has often been argued that from its inception the Community was to be more than a trading union and the Treaty of Rome specifically provided for common policies to be developed in certain economic sectors such as agriculture and transport. However, no specific provision was made for environmental policy. In 1957, this omission was perhaps understandable since there were few international initiatives that had been adopted, and these were mostly concerned with particular types of pollution. There was certainly no coherent environmental policy among the six original Member States (Belgium, France, Germany, Italy, Luxembourg and The Netherlands).

The Paris Summit in 1972 witnessed a significant change in the attitudes of the Community Heads of State, who took the view that economic expansion should result in improvement in the quality of life. That same year also saw a number of major international initiatives including the United Nations Stockholm Conference, which gave rise to the United Nations Environmental Programme (UNEP).

The Head's of State agreed that the Commission should draw up a Community-wide environmental policy. In response to this, the Commission formulated the "First Action Programme for the Environment". The details of this programme and subsequent Environmental Action Programmes are discussed later.

In the EU, the main instrument for implementation of environmental policy is the Directive. EU Directives require the Member States to achieve, within a specified period, the results prescribed within them. The Directives allow Member States to choose the method by which the Directive will be implemented into their legislative systems. In the

United Kingdom, implementation of EU Directives can be by way of primary legislation (Acts of Parliament) or by secondary legislation (statutory instruments), or an order under the European Communities Act 1972, which states that without the need for further legislation, all the provisions of the Treaties governing the Common Market and the rules made under them can be given legal effect in the United Kingdom. Therefore, EU legislation may in certain circumstances be directly effective. If there is a conflict between EU and UK legislation, even if a UK Act of Parliament has been passed, the EU legislation is supreme and takes precedence. Numerous Directives have since been issued by the EU relating to the environment, specific examples of which are discussed below in the relevant sections.

Single European Act 1986

This Act, which provides for the realisation of the internal market (i.e. that within the EU there are no internal frontiers thus ensuring the free movement of capital, goods, services and persons), revises the Treaty of Rome and establishes the objectives, principles and conditions for EU action on the environment. Article 130R sets out the following objectives:

(a) to preserve, protect and improve the quality of the environment;
(b) to contribute towards protecting human health; and
(c) to ensure a prudent and rational utilisation of natural resources.

Article 130S of the Act deals with the decision-making rules adopted by the European Council. In principle, unanimity is still required but the Council can define (by unanimous decision) those matters on which decisions are to be taken by qualified majority.

Article 130T simply enables any Member State to adopt more stringent measures than those adopted by the EU, provided these are compatible with the Single European Act (Side, 1993).

The Maastricht Treaty

The Maastricht Treaty signified the agreement on the texts of the Treaty on European Union and Economic and Monetary Union and associated protocols. The Treaty sets the limits of action by the principle of subsidiarity (i.e. the requirement added by Article 3B of the Maastricht

Treaty that the EU should only make laws that are best made at that level, leaving Member States to take measures that are better taken at a national level), revises the powers of the European Parliament, and provides for a common defence and foreign policy and for joint actions on justice, home affairs and immigration.

In terms of the environment, the Treaty restates the objectives of EU policy aiming for a high level of environmental protection, taking into account the diversity of situations in the various regions of the EU and introduces qualified majority voting except for fiscal matters, town and country planning and land use measures and on matters concerning choice of energy sources (Side, 1993).

Environmental Action Programmes

The Environmental Protection Programmes provide the general framework for the development of EU environmental policy. The First Action Programme (1973–76) established the following objectives (Side, 1993):

(a) to prevent, reduce as far as possible and eliminate pollution and nuisance;
(b) to ensure sound management of, and avoid any exploitation of, resources or of nature which cause significant damage to the ecological balance;
(c) to guide development in accordance with quality requirements especially by improving working conditions and the settings of life;
(d) to ensure that more account is taken of environmental aspects in town planning and land use; and
(e) to seek common solutions to environmental problems with states outside the EU and particularly with international organisations.

The general principles of EU environmental policy are also defined by the First Action Programme. These principles are as follows (Side, 1993):

(1) The best environmental policy consists of preventing the creation of pollution and nuisance at source, rather than subsequently trying to counteract their effects.
(2) Effects on the environment should be taken into account at the earliest possible stage in all the technical planning and decision-making processes.

(3) Any exploitation of natural resources or of nature which causes significant damage to the ecological balance must be avoided. The natural environment represents an asset which can be used but not abused.

(4) The standard of scientific and technical knowledge in the EU should be improved with a view to taking effective action to conserve and improve the environment and to combat pollution and nuisance.

(5) The cost of preventing and eliminating nuisances must in principle be borne by the polluter.

(6) The EU and its Member States must take into account in their environmental policy the interests of developing countries, and must in particular examine any repercussions of the measures contemplated under that policy on the economic development of such countries and on trade with them.

(7) A global environmental policy will be better promoted by a clear, long-term EU environmental policy. The EU should make its voice heard and should make a contribution in appropriate international fora.

(8) Environmental protection is a matter for all, the public should be made aware of its importance. Continuous and detailed educational activity should take place at all levels.

(9) For each different category of pollution it is necessary to establish the appropriate level of protection (local, regional, national, community, international) and the geographical zone to be protected.

(10) National programmes should be co-ordinated at EU level and national policies harmonised without hindering progress at a national level.

The **First Action Programme** set out three specific categories for action (Side, 1993):

(1) Action to reduce and prevent pollution and nuisance, including: the establishment of scientific criteria and methodologies for determining and measuring pollutants and nuisances (with priority to organic halogen compounds, sulphur compounds, nitrogen oxide (NOx), carbon monoxide (CO), mercury (Hg), phenols and hydrocarbons); parameters for determining quality objectives and emission standards; exchange of technical information, research and a documentation system for its dissemination; studies on industrial pollution with a focus on paper and pulp, iron and steel

and titanium dioxide manufacturing; possible actions for dealing with toxic or persistent waste; and a common method of estimating the cost of anti-pollution measures allowing for the application of the "polluter pays" principle.

(2) Action to improve the quality of the environment, including: safeguarding the natural environment, particularly in poor rural areas with regard to farming; examination of scarcity of water resources and exploitation of natural resources; and examination of environmental problems in urban areas. Actions also to be taken to improve the working environment and to promote environmental awareness.

(3) Action in international organisations including a pledge to enter into active co-operation in the area of the environment with appropriate international bodies, particularly the OECD, the Council of Europe and the United Nations.

The **Second Environmental Action Programme** (1977–81) is usually considered as a continuation of the first, reaffirming its objectives and general principles. However, the Second Programme also gave priorities to anti-pollution measures in water and air, as well as specific measures to combat noise pollution.

The **Third, Fourth and Fifth Action Programmes** introduced significant new elements into EU environmental policy and the present character of EU policy is substantially broadened from the initial basic objectives of prevention of pollution, environmental protection and sound management to avoid damaging the ecological balance.

The **Third Programme** (1982–86) highlighted the need to integrate environmental policy with other EU policies and underlined the need to aim for a preventative approach. In addition, specific priorities were established for the development of Europe's environmental policy.

The **Fourth Programme** (1987–92) highlighted the different approaches that might be adopted for pollution control and prevention. Stricter environmental standards were envisaged in the programme and a progressive implementation of an associated environmental education and information policy were stressed. The Programme stressed again the need for integration of environmental policy with other sectors and provided for a number of initiatives in new areas such as biotechnology, and the management of natural resources which includes an integrated approach to the improvement of the coastal zone environment.

The development of the **Fifth Programme** (1993–2000) coincided with the United Nations Conference on Environment and Development in Rio de Janeiro in June 1992, and both the Programme and the premise on which it is based are closely linked to the concept of sustainable development as embraced at the Rio Conference. The Programme stresses the sustainable management of natural resources and provides a more wide-ranging environmental policy than before. It further switches the emphasis away from grouping environmental controls by reference to environmental media such as air, water or land, to looking horizontally at all the environmental implications of various sectors of the economy. In particular, as part of the principle of integrating environmental policies within the EU's other policies, it concentrates on the industry, transport, agriculture, energy and tourism sectors.

A central concept of the Fifth Programme is of shared responsibility, whereby the responsibility for solving environmental problems is shared between government, industry and the consumer. This is an important factor in the development of environmental policy since it changes the emphasis away from legislation and regulation as a means of solving environmental problems and allows market factors such as financial mechanisms or more innovative legal instruments such as civil liability to be applied.

2.3 Principal sources of UK legislation

Environmental Protection Act (EPA) 1990

This Act gives substantial powers to government ministers to issue secondary legislation and sets out the general principles governing the exercise and limits of those powers. The Act operates as a framework for secondary legislation, which in turn is contained within a variety of statutory instruments (usually made up of regulations and orders).

Implementation of legislation is one of the main activities undertaken by the statutory bodies (e.g. the Environment Agency) or a variety of local authorities (i.e. county councils, district councils and unitary authorities). The Government issues guidance notes and circulars which detail how it wishes these bodies to implement their powers. Although the guidance notes do not have the same authority as secondary legislation, they are observed very closely.

The EPA 1990 is divided into nine parts:

Part I   Integrated pollution control (IPC) and local authority air pollution control (LAAPC), later revised into IPPC

Part II   Waste on land
Part III   Statutory nuisances and clean air
Part IV   Litter etc
Part V   Radioactive substances
Part VI   Genetically modified organisms
Part VII   Nature conservation in Great Britain and countryside
             matters in Wales
Part VIII   Miscellaneous
Part IX   General

From the auditor's perspective the most important parts of the Act are Part I (IPC and LAAPC) and Part II (Waste – duty of care, waste management, controlled waste and special waste).

Part I – IPC and LAAPC

Part I of EPA 1990 introduced the system of both integrated pollution control (IPC) and local authority air pollution control (LAAPC). The IPC system was designed to provide a framework for controlling emissions to land, air and water from the most potentially polluting or technologically complex industries and other processes throughout England and Wales. It was introduced progressively from 1 April 1991, with the enforcing authority being the Environment Agency (post 1995, formally being Her Majesty's Inspectorate of Pollution (HMIP)). However, in order to implement the EU Directive 96/61/EC on Integrated Pollution Prevention and Control (hereinafter the IPPC Directive), major changes were needed to UK legislation with the repeal of Part I. This was undertaken by the implementation of the Pollution Prevention and Control Act 1999, via the Pollution Prevention and Control (England) Regulations 1999 and the Pollution Prevention and Control (Scotland) Regulations 1999, which came into force on 31 October 1999. Ultimately, Part I of EPA 1990 will be entirely replaced by the 1999 Act. However, until all the processes are phased into IPPC, Part I will continue to be relevant. For this reason, the following section deals with the individual aspects of both IPC and LAAPC, before going into more detail on the IPPC system.

Integrated pollution control

The main objectives of IPC were to:

(a) prevent or minimise the release of prescribed substances and to render harmless any such substances which are released; and

(b) develop an approach to pollution control that considers discharges from industrial processes to all media in the context of the effect on the environment as a whole.

The Secretary of State exercised his powers to prescribe processes through regulations via the Environmental Protection (Prescribed Processes and Substances) Regulations 1991 (SI No 472), as amended, which are detailed later in this section. Part A of the Regulations outlined those processes which were subject to IPC. Through these Regulations, prescribed substances which in the Secretary of State's view are the most potentially harmful or polluting when released into the environment were also identified for particular control under IPC, and consequently were subject to special requirements to ensure that their release to specified media was prevented, minimised or rendered harmless. IPC is therefore concerned with the control of releases to all three environmental media – air, water and land (as defined in EPA 1990, s 1).

In setting the conditions for an authorisation, the Environment Agency has a duty to ensure that certain objectives are met:

(a) That the Best Available Techniques Not Entailing Excessive Cost (BATNEEC) are used to prevent or, if that is not practicable, to minimise the release of prescribed substances into the medium for which they are prescribed; and to render harmless both any prescribed substances which are released and any other substances which might cause harm if released into any environmental medium;

(b) That releases do not cause, or contribute to, the breach of any Direction given by the Secretary of State to implement European Union international obligations relating to environmental protection, or any statutory environmental quality standards or objectives, or other statutory limits or requirements;

(c) That when a process is likely to involve releases into more than one medium (which will probably be the case in many processes described for IPC), the best practicable environmental option (BPEO) is achieved (i.e. the releases from the process are controlled through the use of BATNEEC so as to have the least effect on the environment as a whole).

Guidance notes have been issued by the Environment Agency to advise on the application of IPC (see Appendix 2). These guidance notes give detailed information about specific individual processes. For IPC, the first guidance notes were issued under the IPR series, with the recently issued second series identified as S2 (note that, because of the implementation of the Pollution Prevention and Control Act 1999, new guidance notes for various sectors will be published in line with the sectorial phasing of certain industries). In setting conditions for IPC authorisations, the regulatory authority is required by EPA 1990 to have regard to the IPR guidance, but it may depart from this and be either stricter or more relaxed if special circumstances justify this.

Within the remit of environmental auditing, assessing a site in terms of complying with the provisions set out in its authorisation may be a principle objective (depending on the audit remit). In addition, when considering issues pertinent to contaminated land (especially important within due diligence auditing), note that any contamination of soil and/or groundwater which was the result of a release from an authorised process under IPC, could be seen as a breach of the authorisation, and hence the site may be subject to regulatory action. The usual remedies would apply (i.e. prosecution or the service of an enforcement notice), with the Environment Agency able to take reasonable steps under section 27 of EPA 1990 towards remedying the harm and recovering the costs where it has been established that the offence is causing harm.

Local authority air pollution control

The main features of the LAAPC provisions within EPA 1990 are that prescribed processes must not be operated without an authorisation from the relevant local authority. Existing processes were allowed to continue operation until being refused or receiving authorisation. In addition, in the event of an application being refused, operation may be continued pending an appeal decision.

One of the underlying principles of EPA 1990 was that of BATNEEC (see p27, below). Under this principle, as with IPC, local authorities are under a statutory obligation to include in any authorisation a requirement for the process to be operated using BATNEEC, so as to prevent and minimise emissions of prescribed substances and to render harmless any substance that may be emitted. Furthermore, in addition to any specific conditions included in an authorisation, all authorisations implicitly impose a duty on the operator of the process to use BATNEEC in relation to any aspect of the process not covered by the specific conditions.

Guidance notes have been issued to local authorities which detail information about specific processes and BATNEEC (see Appendix 2). Although the guidance notes are issued to the local authorities, they give helpful information to those operators that are regulated by LAAPC authorisations.

One of the remaining main features of the LAAPC system is the provision by local authorities of public registers detailing the local air pollution control processes in the area.

Pollution Prevention and Control Act 1999 and related Regulations

It is important to note here the change in the political face of the United Kingdom. Because of the devolution of legislative powers to Scotland and Wales, the national regimes which will implement the Pollution Prevention and Control Act 1999 will develop separately. Regulatory powers for Wales will fall to the National Assembly for Wales (under the Wales Act 1998), with the Scottish Executive responsible for Scotland under the Scotland Act 1998.

One advantage to the United Kingdom in the implementation of IPPC is that the IPPC Directive is principally based on the UK IPC system. The principal result of IPPC will be that the regulators will be required to consider a wider range of environmental impacts before a permit is issued. Under the new IPPC system, noise, vibration, waste minimisation, energy efficiency, the use of raw materials, accident prevention and site restoration (soil and groundwater contamination) will also have to be considered. In addition, the IPPC system will apply to significantly more installations (the Government estimates 7,000 installations will be covered by IPPC) including pig and poultry installations, firms in the metal industry (currently covered by LAAPC) and the food and drink industry. The Pollution Prevention and Control (England) Regulations 1999 and their counterpart Scotland Regulations set out various types of installation.

All processes which previously fell under the IPC regime will be covered by the IPPC system. Therefore, if a process has an IPC authorisation, it will eventually fall under the IPPC regime. These processes are classed as A(1) under the Regulations for England and Wales (to be regulated by the Environment Agency) and Part A under the Scottish Regulations (to be regulated by the Scottish Environmental Protection Agency).

Some LAAPC processes will be subject to the provisions of the IPPC Directive. At present, the regulation of these processes will be undertaken by the local authorities of England and Wales, with the

Environment Agency reserving the right to specify minimum conditions for discharges to controlled waters and sewers. These installations are described as A(2)/Part A installations.

Processes which were formally regulated by LAAPC, but are not included within the IPPC Directive, will be authorised under the PPC regulations as Part B processes, and will continue to be regulated by the local authority air pollution control (England and Wales) and air pollution control by the Scottish Environmental Protection Agency in Scotland.

Another principal distinction from the previous IPC system, is the use of the term "installation" rather than "process". An "installation" is defined by the Directive as "a stationary technical unit, where one or more activities listed in Annex I are carried out, and any other directly associated activities which have a technical connection with the activities carried out on that site and which could have an effect on emissions and pollution". The definition of an installation is wider than that of a process, with the government applying two distinct limbs to the definition:

(a) an installation must be a stationary technical unit, where one or more activities listed in Annex I of the Directive are undertaken, and
(b) any directly associated activities must have a technical connection with the Annex I activity (or activities), and those which could have an effect on emissions are also part of the installation.

As with IPC and LAAPC, Guidance Notes (Best Available Technique Reference documents (BREFnotes)) will be published by the Government with the objective of providing clear, indicative performance standards for both new and existing plant and upgrading timetables for the latter. Unlike the former IPC system, there are no prescribed substances within the Directive, but rather an indicative (non-exhaustive) list of pollutants for which emission limits may be set.

One of the principal areas of discussion with respect to the new system is the requirement for a site condition report to be prepared and submitted for each application. The discussion revolves around the standard to which investigations (and eventual remediation) would need to be applied. The contaminated land regime introduced under the Environment Act 1995 applies the concept of "suitability for use" (see Chap 8). However, under the IPPC system no further significant degradation of the soil and groundwater is accepted. This principle stems from the view that site contamination at an IPPC installation should not arise in the first place. However, if it does, then the reference point for the remediation of the site should be the condition of the site prior to the pollution incident occurring and not the considered future usage of the site. On this basis, the content of the original site condition report is important, since it will be from this report that any presumed contamination at application surrender will be referenced against. At the time of writing the Environment Agency is expected to be producing guidance on the characterisation and assessment of sites, promoting a consistent approach to site characterisation.

BATNEEC v best available techniques

One of the underlying principles of EPA 1990 and the associated IPC regime was the concept of BATNEEC. The term BATNEEC gained increasing exposure in international legislation and agreements relating to environmental protection, most notably the EU Air Framework Directive (84/360) and Dangerous Substances Directive (76/360). The EU uses the term "best available technology", whereas EPA 1990 used the term "best available techniques". The term "techniques" was intended to include technology, but in addition to hardware it is also intended to include operational factors.

All the IPC-prescribed processes contained within Part I of EPA 1990 were subject to BATNEEC requirements. For each individual process it was the role of the Environment Agency to establish what constitutes BATNEEC (subject to appeal to the Secretary of State). Guidance on what constitutes BATNEEC for each individual process is contained within the Chief Inspector's Guidance Notes. However, this concept was rejected within the IPPC system and has now been replaced with best available technique (BAT). The conditions on permits now make reference to BAT and, as detailed in the Directive, the UK 1999 Regulations (Sched 2) list matters to which special consideration is to be given. Guidance on what will constitute BAT will be issued in the form of BREFnotes. Although the caveat "not exceeding excessive cost" has been removed from the IPPC system, the Government has stated that it would not expect any extra costs to be incurred by operators in the switch from BATNEEC to BAT. For this reason, the full description of the principles underlying BATNEEC are described below.

Although "BATNEEC" is not defined in EPA 1990, the DoE's guide to IPC clarifies the definition of the terms incorporated within the BATNEEC principle. The term "techniques" is defined in section 7(10) of EPA 1990 and embraces both the process and how the process is operated. It therefore should be taken to include the design of the process, the components of which it is made, the manner in which it is connected together, the numbers and qualifications of the staff, working methods, training, supervision and the design, construction and maintenance of the buildings.

The term "available" should be taken to mean procurable by the operator of the process in question. It does not imply that the technique should be in general use, but it does require general acceptability. It therefore includes a technique which has been developed (or proven) at a scale which allows implementation in the relevant industry with commercial business confidence. It does not imply that techniques outside the United Kingdom are unavailable.

The term "best" must be taken to mean the most effective in preventing, minimising or rendering harmless polluting emissions. As there may be one or more techniques that achieve comparable results, there may be more than one best technique.

The DoE's guide to IPC, published in 1991, states that "not exceeding excessive cost" needs to be taken in two contexts: new processes or existing processes.

For new processes, the best available technique will be used, but in the Secretary of State's view the presumption can be modified by economic considerations, where the cost of applying the best available technique would be excessive in relation to the nature of the industry and the environmental protection achieved. The revised version of the guide published in January 1993 emphasised that for new processes it would be the economic condition of the industrial sector, rather than of that a specific company, that would determine what was BATNEEC. The following principles are also clarified in the revised guidance:

(a) the cost of BAT will be weighted against the environmental damage associated with the process (*i.e.* the greater the environmental damage induced, the greater the allowable costs of BAT will be before they can be classified as excessive);
(b) even after applying BATNEEC, if serious harm is still being applied to the environment, an application may be refused;
(c) an objective approach to BATNEEC is needed (*i.e.* in the assessment of the costs of BATNEEC); however, the lack of profitability of a particular business should not be a determining factor.

For existing processes, the DoE's (now encompassed within the DETR) original guidelines stated that the inspector will be concerned over the time scale by which old processes are to be upgraded to new standards or decommissioned. This statement is correspondingly revised in the latest guidelines, which refer to new standards or as near to new standards as possible. The DoE's (DETR's) general approach is based around the approach adopted by the EU Air Framework Directive (84/360).

Article 12 of the Directive requires, where necessary, the imposition of appropriate conditions of authorisations for the Directive on the basis of the development of BAT and the environmental situation and also on the basis of avoiding excessive implementation costs for the plant in question. Under Article 13, which only applies to processes operational prior to July 1987:

> "in the light of an examination of developments as regards best available technology and the environmental situation, the Member States shall implement policies and strategies, including appropriate measures, for the gradual adaptation of (specified) existing plants to the best available technology, taking into account in particular:
> – The plant's technical characteristics;
> – Its rate of utilisation of its remaining life;
> – The nature and volume of polluting emissions from it;
> – The desirability of not entailing excessive costs for the plant concerned, having regard in particular to the economic situation of undertakings belonging to the category in question."

"BATNEEC" can therefore be expressed in technological terms (*i.e.* a requirement to employ specific hardware), but it may also be expressed in terms of emission standards. Having identified the best techniques and the relative release levels associated with those techniques, it is therefore possible to express BATNEEC as a performance standard.

Best practicable environmental option (BPEO)

The Royal Commission on Environmental Pollution discussed the concept of BPEO in its 1988 report. For a given set of objectives the BPEO is the option that provides the most benefit or least damage to the environment as a whole, at acceptable cost, in the long term as well as in the short term. "BPEO", as with BATNEEC, is not defined in EPA 1990; rather, the Act states that BATNEEC will be used for minimising the pollution which may be caused to the environment taken as a whole by the releases having regard to the BPEO available with respect to the substances that may be released. Furthermore, the Act's requirement only applies where the process is likely to involve the release of substances into more than one environmental medium, and does not allow for a wider interpretation of BPEO, such as considering the fundamental need for the process, or extending the remit along the supply chain to consider the choice of raw materials.

The 1991 DoE guide to IPC recognised that a full environmental impact statement for each process option was not practicable and described a scenario in which assessment would be undertaken in the main areas where the process was likely to impact on the environment (whether globally, regionally or locally). This statement was qualified by the use of an example where a particular local environment issue or sensitivity exists, in which case the operator would need to demonstrate that his/her proposed process/technique takes adequate account of it. However, in the revised 1993 guidance, this statement was withdrawn and no further view was offered.

Environmental Protection (Prescribed Processes and Substances) Regulations 1991

These Regulations, which came into force in England and Wales on 1 April 1991 and in Scotland on 1 April 1992, are due to be repealed by the Pollution Prevention and Control Act 1999. The 1991 Regulations include both the processes and substances to which IPC initially applied. The list of prescribed processes is contained within Schedule 1 to the Regulations, which in turn is divided into Part A and Part B processes. Part A processes are those prescribed for IPC, whereas Part B processes are those prescribed for LAAPC, as described above.

Part II – waste management

The Control of Pollution Act 1974 made it an offence to cause or knowingly permit waste to be deposited without a waste disposal licence. Part II of EPA 1990 details a number of provisions relating to waste management which are targeted at overcoming a number of difficulties thrown up by the 1974 Act. These expanded on the waste disposal licensing provisions so as to require a waste management licence for operations including the storage and handling of waste as well as its disposal, and state explicitly that a breach of a licence condition is an offence in itself. In addition, under EPA 1990 it is an offence to knowingly cause or knowingly permit waste to be deposited unless duly licensed; to keep, treat or dispose of waste in a manner likely to cause pollution of the environment, or harm to human health (s 33(1)(c)); and defines "pollution of the environment" (s 29(3)) as:

> "Pollution ... due to the release or escape (into any environmental medium) from:

(a)   the land on which controlled waste is treated;

(b)   the land on which controlled waste is kept;

(c)   the land in or on which controlled waste is deposited;

(d)   fixed plant by means of which controlled waste is treated, kept or disposed of, substances or articles constituting or resulting from the waste and capable (by reason of the quantity or concentrations involved) of causing harm to man or any other living organisms supported by the environment."

However, implementation of Part II of EPA 1990 experienced a number of problems. As previously mentioned, the Act is set out as a framework and correspondingly no time scale was contained within the Act. As a result, there was a series of delays which were compounded by the Government's decision to create the Environment Agency, which took over the waste regulation functions previously exercised by local authorities.

Part II requires that, prior to a licence being surrendered, the Environment Agency or the Scottish Environmental Protection Agency must be satisfied that the condition of the site is such that it is unlikely to cause pollution of the environment, or harm to human health. It is important to note that the new contaminated land regime (see Chap 8) does not apply to sites with a current waste management licence.

Duty of care

Part II introduced a new concept to waste management, that of the duty of care with regard to waste. Section 34(1) states that it:

"shall be the duty of any person who imports, produces, carries, keeps or disposes of controlled waste, or as a broker, has control of such waste, to take all such measures applicable to him/her in that capacity to ensure that the waste is disposed of safely, i.e. to prevent the escape of the waste from his/her control or that of any other person; that the transfer is only to an authorised person or to a person for authorised transport purposes and that there is transferred such a written description of the waste as will enable other persons to avoid a contravention and to comply with the duty ... as regards the escape of waste."

The only exception to the duty of care is for domestic occupiers regarding their own household waste. The Act therefore attempts to induce a duty of care from cradle to grave, therefore extending the responsibility to the producer, who will bear the primary responsibility for providing a sufficient description of the waste so that inappropriate disposal options are not undertaken, as well as being responsible for checking the registration documentation of any carrier before contractual arrangements are undertaken and conducting regular checks of the documentation.

The transfer of waste must only be undertaken by an authorised person, which includes waste collection authorities in England and Wales (or waste disposal authorities in Scotland), the holders of waste management licences (or those who are exempted from having to hold one) and registered carriers of controlled waste (or those exempted from having to register). A system of carrier registration, which was introduced by the Control of Pollution (Amendment) Act 1989, was implemented by the Controlled Waste (Registration of Carriers and Seizure of Vehicles) Regulations 1991 (SI No 1624).

The duty of care is supported by a statutory Code of Practice which was published by the DoE in 1991. The Code provides advice to waste holders on how to comply with the duty of care and discharge their duties.

The Environmental Protection (Duty of Care) Regulations 1991 (SI No 2839) came into force in April 1992 and impose responsibilities for record-keeping on the transferor and transferee of waste. The Regulations require both parties to keep complete records of transfer notes containing information relating to the transfer of waste between them. Although there is no prescribed format for what should constitute a transfer note, the note should be signed by both parties and its contents should include the following:

- A description of the waste
- The quantity of waste being transferred
- Type of container
- Name, address and capacity of the person transferring the waste (*i.e.* whether producer, importer or carrier)
- Name, address and capacity of the person receiving the waste
- Reference to necessary legislation including licence number
- Name of the relevant waste regulation authority
- Date and place of transfer.

A suggested layout for a Controlled Waste Transfer Note is given in the Code of Practice. Although a breach of the duty of care does not in itself create a civil liability, it is a criminal offence and correspondingly is subject to a penalty fine. Furthermore, such a breach may provide evidence of negligence in the event that a third party suffers damage as a result of the escape of waste.

"Waste" is defined in section 75(2) of EPA 1990 as including:

"(a) any substance which constitutes a scrap material or an effluent or other unwanted surplus substance arising from the application of any process; and
(b) any substance or article which requires to be disposed of as being broken, worn out, contaminated or otherwise spoiled."

In general, whether something is waste is measured by the attitude of the person disposing of it, not the recipient. Therefore, materials that are sent for recycling are considered as waste under the Act and are therefore the subject of any regulations made under it, and will remain waste until they are recycled. Section 75(4) defines controlled waste as household, industrial or commercial waste. The Controlled Waste Regulations 1992 (SI No 588), which came into force in April 1992, determine what is and what is not controlled waste, and break down what is controlled waste into its constituent parts (*i.e.* household, industrial and commercial).

Special waste is a type of controlled waste which is considered to be sufficiently dangerous to treat, keep or dispose of that special provisions are required. The Special Waste Regulations 1996, which came into force on 1 September 1996, specify what constitutes special waste and provide a list of special wastes detailed in Schedule 2 and a detailed classification system for determining whether a waste is special. Disposal of special waste requires pre-notification of the proposed waste to the relevant waste disposal authority, and utilises consignment notes, which ensure that the transfer of such waste is properly documented. The revised Regulations also include in Schedule 1, Part I what form a consignment note should take.

Statutory nuisance

Part III of EPA 1990 deals with the concept of statutory nuisances, which derives from the civil tort of nuisance. In essence this requires the occupier of land to behave reasonably towards his neighbours. Section 79 of the Act details specific statutory nuisances, and for these a specific procedure is provided by the statute which the relevant local authority, or any individual, may invoke to abate and stop the activity if they are aggrieved by its operation. Provided that activities are prejudicial to health, activities that can be controlled in this way are those producing:

- Smoke, fumes or gases
- Any dust, steam, smell or other effluvia arising on industrial trade or business premises
- Any deposit or accumulation
- Noise.

It should be pointed out at this stage that there is no reason why contaminated land may not fall under this statutory control where it can be proven that the premises are "in such a state as to be prejudicial to health or a nuisance", or there is an "accumulation or deposit which is prejudicial to health or a nuisance". However, this application will be disapplied upon the implementation of the contaminated land regime.

A local authority may serve an abatement notice which details the steps needed to be undertaken to stop the nuisance, and a date by which these steps must be undertaken before proceedings can be initiated through a magistrates' court. Alternatively, an individual may seek sanction directly from a magistrates' court for an appropriate order which is comparable to an abatement order. In many cases, where the nuisance is resulting from a business or commercial activity, it is a defence to argue that the best practicable means have been used to avoid the nuisance being created.

Water legislation

Water Act 1989 and Water Resources Act 1991

Water legislation in the United Kingdom was significantly altered by the Government's policy of privatisation and the resulting Water Act in 1989. This Act created the National Rivers Authority (NRA) and enhanced the system of consents previously introduced by the Control of Pollution Act 1974, which extended the geographical coverage of the controls for most discharges to include the territorial sea (up to a three-mile limit), inland waterways and groundwater. Important changes with the 1989 Act are the introduction of statutory water quality standards and the introduction of a system of charging for trade and sewerage discharges.

The Water Act 1989 was consolidated in the Water Resources Act 1991, which limited the NRA's responsibility for total control over discharges to coastal and inland waters through the development of HMIP, which in turn had responsibility for IPC.

Under section 85(1) of the 1991 Act, a person commits an offence if he causes or knowingly permits a poisonous, noxious or polluting matter or any solid matter to enter any controlled waters. Controlled waters include relevant territorial waters, some coastal waters, inland waterways and groundwater. Furthermore, subsections of section 85 create similar offences in respect of the discharge of trade effluents or sewerage effluent, and impeding the proper flow of any inland fresh waters in a manner leading or likely to lead to a substantial aggravation of pollution owing to other causes or the consequences of such pollution.

This legislation correspondingly forms the basis for a system of licensing for discharges to controlled waters (but not sewers and drains leading to sewerage discharge plants). Discharges to sewers are controlled under section 118 of the Water Industry Act 1991, which allows for the discharge of trade effluent to public sewers with the consent of the relevant sewerage undertaker. Discharges to controlled waters are under the control of the Environment Agency (formally the NRA). It is an offence to discharge without the appropriate consent and there are a number of materials, such as petrol which may not be discharged to sewers. In addition, none of the 23 "red list" substances may be discharged to sewers, except with the Secretary of State's consent. The red list contains the most toxic materials, including mercury, cadmium compounds, organochlorine pesticides and polychlorinated biphenyls (PCBs).

Groundwater Regulations 1998

The Groundwater Regulations 1998 came into force on 1 January 1999 and are designed to help prevent pollution of groundwater by controlling discharges or disposal of certain dangerous substances, if not already covered by existing legislation. The emphasis of the Regulations is to prevent the direct and indirect discharges of List I substances to groundwater and to control pollution resulting from the direct or indirect discharge of List II substances. The Environment Agency is tasked with the responsibility of ensuring compliance with the Regulations and issuing authorisations for the disposal of listed substances (including materials which contain these substances) onto or into land.

List 1 substances are the most toxic and must be prevented from entering the groundwater. These include pesticides, sheep dip (organophosphates), solvents, hydrocarbons, mercury, cadmium and cyanide. No authorisation can be granted which would permit the direct discharge of any List I substance. In addition, an authorisation cannot be granted for any activity which may result in the indirect discharge of a List I substance, unless the activity has been the subject of prior investigation.

List II substances are less dangerous than those specified in List I, although if they are disposed of in large quantities they could be harmful to groundwater. Substances specified in List II include some heavy metals, ammonia and phosphorus. No authorisation resulting in any direct or indirect discharge of List II substances can be granted without prior investigation.

All activities requiring an authorisation must first be subject to prior investigation, which in turn will allow assessment of whether the activity is acceptable. In most cases it is anticipated that the information supplied, combined with any additional information held by the Agency, will be sufficient to allow a decision on the activity to be made. However, allowances are made so that in certain circumstances a visit to the site by Agency personnel may be required, and where the proposed activity is complicated further investigations may also be required, such as intrusive works.

The regulations define "groundwater" as "any water contained in the ground below the water table". The Regulations do not allow for any entry of List I substances into the groundwater unless the groundwater has been classified as being permanently unsuitable for other uses. Where there is a risk of pollution of the groundwater, the Agency may serve a Regulation 19 Notice prohibiting the activity, or allowing it to continue subject to certain provisions. Statutory Codes of Practice will in time be made available by the Agency, which will detail those activities which may result in an unintentional release of a listed substance. Such activities which may be subject to a statutory code include the manufacture, storage or use of a listed substance, such as a chemical works, oil storage or distribution depots.

Authorisations are only required where a disposal is not already covered by an IPC (soon to be IPPC) authorisation, or a waste management licence, or by a discharge consent under the Water Resources Act 1991.

Anti-Pollution Works Regulations 1999

Introduced by the Environment Act 1995, the implementation of these Regulations gives the Environment Agency a significant new enforcement option. An Anti-Pollution Works Notice, if served, requires the person served to carry out specified works (or actions) to deal with pollution of controlled waters.

A notice can be served on any person who caused or knowingly permitted poisonous, noxious or polluting matter or solid waste to be present in a place, from which it is likely in the Environment Agency's opinion to enter controlled waters. Similarly, a notice can be served on any person who caused or knowingly permitted the matter to be present in a place from which it might be released. The Agency has stated that a works notice may not always be applied, but rather would be one of the remedies considered by the Agency. It is also important to note that unlike the contaminated land provisions, there is no reasonable excuse built into the legislation for failing to comply with a works notice, although there are appeal procedures which can be followed.

Hazardous substances

A new system of hazardous substances consents was introduced with the Hazardous Substances Consents – The Planning (Hazardous Substances) Act 1990, which came into force on 1 June 1992. The Act makes it an offence to have present on any site in England and Wales any one or more of the 79 listed hazardous substances, contained within the Planning (Hazardous Substances) Regulations 1992 (SI No 656), in amounts equal to or greater than the controlled quantity, prescribed for each substance, unless there is an appropriate hazardous substances consent. The Town and Country Planning (Hazardous Substances) (Scotland) Regulations 1993 (SI No 323 (S 31)) introduced similar provisions for Scotland. The lists of hazardous substances are broken down into three groups: toxic substances, highly reactive and explosive substances, and flammable substances which are not within either of the first two categories.

The majority of the substances listed are the same as those indexed under the Notification of Installations Handling Hazardous Substances Regulations 1982 (SI No 1357) and the Control of Industrial Major Accident Hazardous Substances Regulations 1984 (SI No 1902, as amended). The new legislation does not apply to controlled waste, radioactive waste or explosives, as they are already controlled under different legislation.

The 1982 Regulations specify dangerous substances and quantities which trigger an obligation to inform the Health & Safety Executive

(HSE) three months before the person proposing to undertake the activity involving these substances actually undertakes the activity. The substances listed in the Regulations are divided specifically into named substances and groups of hazardous substances which can encompass a number of different substances. However, these Regulations are now effectively redundant because of the introduction of the Planning (Hazardous Substances) Act 1990 controls.

The 1984 Regulations apply where a person has control of an industrial activity where there are substances present in sufficient quantities to be capable of producing a major accident hazard (note that all 1984 Regulations sites are also 1982 Regulations sites, but not vice versa). Duties differ according to the quantity of the dangerous substance stored at a particular site. In general there are two limit values placed on each individual substance: the first requires a report to the HSE where a substance exceeds the higher limit value for that substance, and the second outlines a duty to report safe operation where the lower limit is exceeded. In addition, other duties relating to the regulations include, but are not limited to:

(a) the preparation and maintenance of an up-to-date on-site emergency plan detailing how a major accident would be dealt with; and
(b) the preparation and keeping of an up-to-date off-site emergency plan by the local authority (county council).

The Control of Major Accident Hazard Regulations 1999 replace the 1984 Regulations, although broadly following the same principles. The principle change relates to the inclusion of the Environment Agency and its Scottish counterpart as joint competent authorities with the HSE. Additional changes include the use of a smaller list of named substances with generic categories (*e.g.* toxic) for non-listed substances, a duty on companies to prepare a major accident prevention policy (particular emphasis on safety management systems), and a requirement for the safety report to approve the construction-related safety measures for a new plant before construction of the new plant can go ahead (only required for sites with larger inventories of hazardous substances).

Control of substances hazardous to health

The Control of Substances Hazardous to Health Regulations 1988 took effect on 1 October 1989. They were subsequently amended in 1994,

1996, 1997 and 1998. These Regulations apply to virtually all work activities wherever substances which are potentially hazardous to health are used or produced. It is an offence to fail to comply with their requirements and penalties may be incurred under the Health and Safety at Work Act 1974. The principal objectives of the Regulations are to:

(a) provide one set of regulations covering substances hazardous to health as defined, including the many substances not specifically covered by any existing provisions, together with processes which are at present covered only where they occur in factories;
(b) set out those principles of occupational health including those of occupational medicine and hygiene which should be followed; and
(c) make provisions for any future changes in standards of control necessary as a result of the discovery of hitherto unsuspected substances and to encourage the use of new technology, including new techniques for the control of exposure.

Although the Regulations are designed to apply to the whole spectrum of industry and commerce, there are a number of industries and trades for which they are especially applicable:

(a) major manufacturers and bulk users of chemicals;
(b) users of substances in circumstances most likely to involve high exposure levels;
(c) users or handlers of substances/intermediaries which are highly toxic or present special risks;
(d) dusty trades including ceramic and refractory industries, quarrying, foundries and metal manufacturing/finishing processes; and
(e) users of processes which generate substances hazardous to health in appreciable quantities (e.g. certain welding, cutting, grinding, milling or sieving operations).

The Regulations are divided into 19 regulations with nine associated schedules. Each regulation has a General Approved Code of Practice which provides practical guidance to these provisions. Although failure to comply with any provisions laid down in each Code is not in itself an offence, that failure may be taken by a court in criminal proceedings as proof that a person has contravened the Regulations to which the provision relates. Such a case would require the person concerned to satisfy a court that he had complied with the Regulations in some other way.

The essential requirement of the Regulations is for an assessment to be made of any health risks associated with the work being undertaken (reg 6) and details concerning the measures that need to be taken to protect peoples health and meet the requirements of other regulations. The recommended procedures underlying the Regulations follow an established pattern:

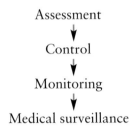

Assessment
↓
Control
↓
Monitoring
↓
Medical surveillance

The employer must provide information, instruction and training to employees relating to the risks and to the precautions that should be taken and, where an accurate assessment has been undertaken, an employer should be able to identify the necessary measures to control the risks.

In order to comply with the 1988 Regulations, the organisation concerned will be involved with five key processes:

(1) Gathering information about the substances, the work and the working practices, including how the substances are used.
(2) Evaluating the risks to health.
(3) Deciding what further precautions may be required to comply with the legal requirements under the 1988 Regulations.
(4) Recording the assessment, unless clearly unnecessary.
(5) Deciding when the assessment needs to be reviewed.

Note that poor control can create a substantial risk, even from a substance with a low hazard rating. With correctly applied precautions, the risk of being harmed by even the most hazardous substance can be adequately controlled.

The responsibility with respect to these Regulations is placed firmly with the employer. It is the employer who is responsible for the assessment and it is the employer who is required to check which hazardous substances are or might be present in the workplace, how they can be recognised and in what form they are present.

Environment Act 1995

Apart from introducing the Environment Agency and the Scottish Environmental Protection Agency, the Environment Act 1995 mainly amended previous Acts, particularly EPA 1990.

Part I of the 1995 Act introduced two new environment agencies:

(a) the Environment Agency which covers England and Wales and carries out the functions previously undertaken by the NRA, the waste regulation authorities and HMIP; and
(b) the Scottish Environmental Protection Agency which now carries out the functions of the river purification authorities, waste regulation and disposal authorities, HM Industrial Pollution Inspectorate and local authorities in relation to air pollution.

Part II of the Environment Act 1995 amended EPA 1990 by adding a new Part IIA which comprises of 26 new sections providing a statutory framework for contaminated land. Within the scope of environmental auditing (specifically due diligence auditing), the issues surrounding contaminated land are extensive. On this basis, contaminated land is discussed in detail later in this section.

Part III of the 1995 Act relates to National Parks, while Part IV introduces a national air quality strategy. Section 80 requires the Secretary of State to draw up a national strategy, setting out the Government's policy regarding the assessment and management of air quality. This includes strategies on the implementation of EU legislation and international agreements.

Section 82 requires local authorities to carry out occasional air quality reviews for their area. They must complete an assessment of the standards of air quality and whether the air quality objectives are being achieved. In the event of the objectives not being met, section 83 requires the local authority to designate the area as an "air quality management area". Section 84 requires the local authority to prepare an action plan for achieving those objectives in the designated area.

Section 92 of the Act introduces a national waste strategy by adding to EPA 1990 new section 44A, which requires the Secretary of State to draw up a national waste strategy which:

(a) contains a policy statement;
(b) covers the types, quantity and origin of the waste to be recovered or disposed of;

(c) sets out general technical requirements; and
(d) describes any special requirements for a particular waste.

A summary of legislation can be found in *Croner's Environmental Policy and Procedures 1996*.

EPA 1990: Part IIA – contaminated land

A new contaminated land regime will be introduced as Part IIA of EPA 1990. These provisions are supported by a number of statutory guidance documents which in turn discuss the principal areas of interest such as when land is considered to be contaminated, levels of clean up and responsibility for clean up. At the time of writing, the government documents on the implementation of Part IIA is anticipated to take effect as of 1 April 2000. The following is based largely on the information presented in the draft consultation paper issued in September 1999 by the DETR and is a summary of the main principles underlying the forthcoming regime. Therefore, note that any interpretation offered is based on this draft guidance and is given subject to what the authors consider reasonable. The reader is advised to consult the final guidance, in addition to the wide range of legal papers available on this subject, in order to gain a full understanding on the new contaminated land regime.

Although the definition of contaminated land has been the subject of continued debate, section 78A of EPA 1990 defines contaminated land (for the purpose of Part IIA) as:

> "any land which appears to the local authority in whose area it is situated to be in such a condition by reason of substances in, on or under it that:
> (a) significant harm is being caused, or there is a significant possibility of such harm being caused; or
> (b) pollution of controlled waters is being, or is likely to be caused."

The definition is specifically drafted to reflect the intended role of Part IIA in the new contaminated land regime (*i.e.* the identification and remediation of land which is causing an unacceptable risk to human health or the wider environment). On this basis, not all contaminated land may be included within the scope of Part IIA, although it may be included within other statutory regimes. Note that the new contaminated land regime does not apply to radioactive substances.

The risks associated with contaminated land have gained increasing exposure over recent years. The main legal risks associated with contaminated land can be summarised as follows (*Environmental Compliance Manual 1999*):

(a) potential civil liability caused by pollution migrating off-site;
(b) potential remediation costs pursuant to the service of a statutory notice (including the costs associated with investigation, clean up and follow on costs);
(c) potential criminal liability for breaches of legislation (especially in relation to water pollution) resulting from pollution migrating off-site;
(d) planning conditions or obligations associated with redevelopment requiring investigation, restoration or aftercare, which act as a constraint on the scope of development or involve expenditure. Certainly contamination is a material consideration when assessing the development potential of a site and local authorities are now including standard conditions within planning authorisations for the assessment of contaminated land to have been undertaken prior to development.
(e) valuation issues (*i.e.* reduction in the property's open-market value).

Local authority duties under contaminated land regime

Under the new contaminated land regime local authorities are required to inspect their area to identify contaminated land. One of the critical concepts underlying the new regime is that of "suitable for use", with the specific exception applying to where an environmental licence or permit has been breached. In this instance, complete removal of the contamination by the polluter would be required. It is intended that remediation would only be required on those sites which, through a process of risk assessment, the contamination is found to pose an unacceptable risk to health or the environment, taking into account the use of the land.

For a site to be considered as being contaminated, there is essentially a two limb test which must first be satisfied. Section 78B(1) of the Environment Act states that land would be considered to be contaminated by reason of substances in, on or under it when either:

(a) significant harm is being caused or there is a significant possibility of such harm being caused, *i.e.*:
    (i) death, ill health or damage to humans, livestock or crops;
    (ii) structural failure or substantial damage to buildings; or
    (iii) adverse change to a natural habitat protected under UK or EU law; or

(b) pollution of controlled waters is being caused or is likely to be caused, irrespective of whether any harm is caused (note, however, that unnecessarily expensive remediation measures would not be reasonable and cannot therefore be imposed by a remediation notice).

The principal component to be considered when making the determination of whether a site is contaminated is the assessment of the pollutant linkage. The use of a source–pathway–target analysis forms the basis for the assessment of the pollutant linkage. The concepts of contaminant (source), pathway and receptor (target) are defined in the contaminated land guidance as follows:

> "A contaminant is a substance which is in, on, or under the land and which has the potential to cause harm or to cause pollution of controlled waters;
>
> A receptor is either:
> (a)   a living organism, a group of living organisms, an ecological system or a piece of property which
>
>> (i) is a category listed in Table A [see Appendix 6] as a type of receptor, and
>> (ii) is being, or could be, harmed, by a contaminant; or
>
> (b)   controlled waters which are being, or could be, polluted by a contaminant.
>
> A pathway is one or more routes or means by, or through which a receptor:
>> (i) is being exposed to, or affected by, a contaminant, or
>> (ii) could be so exposed or affected."

If when inspecting its area, a local authority identifies a site as contaminated by virtue of a pollutant linkage being established, the authority will then have to establish whether the site should be designated as a "special site". If it is designated as a special site, then the Environment Agency will assume regulatory control over it. The Contaminated Land (England) Regulations 1999 set out the descriptions of contaminated land which may be designated as a special site (see Appendix 6).

The next stage in the procedure is for the local authority and, in the case of a special site, the Environment Agency to establish whether urgent action is required. In this event, the relevant authority has powers to take action itself, rather than proceeding with a remediation notice.

One of the areas of key concern within the property and financial sectors is the method by which the local authority will identify the appropriate person(s) who will bear the responsibility for the remediation of any contaminated land. The history of any given site is often very complex, with numerous occupiers and different industries (all using different substances), all having used the site and subsequently contributing to various degrees to the problems encountered on the site, which in turn have led to the site being designated as contaminated. Correspondingly, various remedial actions may be required, which in turn may not "correlate neatly" with those who will bear the responsibility for the costs. The determination of liability associated with contaminated land is therefore very complex. Although the reader is directed to the guidance for full details as to the procedure for determining liabilities, the following summarises the key points to be considered when considering contaminated land within the due diligence process.

Determining liabilities – procedure

The Class A person (or persons) is the person who caused or knowingly permitted the contaminating substances to be in, on or under the land. Under the definition of a Class A person provided by section 78(F) of Part IIA, the question of liability must be considered separately for each significant pollutant linkage.

The Class B person arises in cases where, after reasonable enquiry, it is not possible to find the Class A person, either for all or any of the pollutant linkages identified at a site. The Class B person is the owner or occupier of the land for the period in question. Note at this stage that there is more limited liability in the case of a Class B person, as opposed to a Class A person principally because of limitations which exist in relation to the pollution of controlled waters (*i.e.* for any given single pollutant linkage, if a Class A person cannot be found and the pollutant linkage relates solely to the pollution of controlled waters, then no liability group would exist and the linkage would be treated as an "orphan linkage"). It is also important to note that the owner or occupier of the land may also be a Class A person by virtue of his own past acts or omissions.

The guidance defines "found" as bearing the same definition as that provided in the *Oxford English Dictionary* and states that the Class A person needs to be in existence to be found (*i.e.* the person is not dead, and/or the company has not been dissolved).

The guidance states that it is ultimately for the courts to decide the meaning of "caused and knowingly permitted" within the context of Part IIA. The guidance does draw a distinction in the example where a person is informed of the presence of a pollutant and then might be considered to have knowingly permitted its presence. The guidance states that this test would only be met where the person had the ability to take steps to prevent or remove that presence and had a reasonable opportunity to do so. This does not mean that when the owner or occupier of the land is notified that the land has been identified as contaminated, they then have the resultant knowledge to trigger the "knowingly permit" test.

There are eight distinct stages in the procedure for determining the appropriate persons who should bear the responsibility for remediation.

Stage I – identifying potential appropriate persons and liability groups. The authority would already have provisionally identified and notified apparent appropriate persons during the course of its earlier inspections. However, at this stage, the authority would need to re-examine this question and establish whether some or all of these persons remain appropriate persons or whether some other persons may now appear to be appropriate persons and notify them accordingly. Exemptions from liability at this stage apply (draft guidance, 1999) where:

(a) the Class A person is exempted from responsibility from liability arising to water pollution from an abandoned mine (s 78J(3)); or
(b) a Class B person is exempted from liability arising from the escape of a pollutant from a piece of land to other land (s 78K); or
(c) a person is exempted from liability by virtue of his being a person acting in a relevant capacity, such as acting as an insolvency practitioner (s 78X(4)).

Note that if all the persons identified within the liability group are removed owing to benefiting from one or more of these exemptions, then the pollutant linkage would be treated as an orphan linkage.

Stage II – characterising remediation actions. This stage relates to where there is more than one pollutant linkage, and hence there is a possibility of one or more remedial actions being required, raising the question of which appropriate person has liability for which remedial action (this in turn is dealt with at Stage III). See the Government guidance for further information in this respect.

Stage III – attributing responsibility to liability groups. In the event that there is only one pollutant linkage, the liability group would bear the costs of the full work being undertaken. If there is more than one pollutant linkage, then the costs would be attributed to each of the liability groups, based on the identification of the required works as set out in Stage II for each pollutant linkage.

Stage IV – excluding members of a liability group. This is an important stage, since it sets out the exemptions which will apply for each Class A and Class B liability group with two or more members. The guidance gives detailed descriptions on six exclusion tests applicable to Class A persons. However, note that these tests cannot be applied such that they exclude all members of the group. There must be at least one person left to which the responsibility can be placed. The exclusion tests are very detailed – see the guidance for further information in this respect. However, the exclusion test which is envisaged to be most widely encountered relates to the "sold with information" exclusion (test 3). This exclusion allows the seller of a piece of land who has caused or knowingly permitted the presence of a significant pollutant in, on or under the land to be excluded from liability by ensuring that the purchaser/tenant of the land (who must also be within the liability group) had full information that would reasonably allow that person to be aware of the presence on the land of the pollution, and its significance within a pollutant linkage, and the seller did not misrepresent the implications of its presence. This is a very important exemption when dealing with property transactions; see the guidance for further information in this respect.

Stage V – apportioning liability between members of a liability group. This stage identifies the protocol by which costs will be attributed to each liability group once all exemptions have been applied.

Stage VI – consultation as to what remediation is to be undertaken. The relevant authority (either the local authority or the Environment Agency) must serve notice on the owner and occupier of the land, in addition to every person identified in the liability group, that the land has been identified as being contaminated. Reasonable endeavour must then be undertaken to consult the relevant parties on the type of remediation that is to be undertaken. The consultation period must be for a minimum period of three months.

Stage VII – restrictions on the service of a remediation notice. Note at this point that there are certain restrictions in which the statutory authority is prevented from serving a remediation notice where:

(a) voluntary remediation is going to take place and the local authority is satisfied that this is the case;
(b) the local authority is the appropriate person; or
(c) the local authority has the power to take emergency action.

In addition, note that there are limits on what remedial actions may be required. When requiring remedial works to be undertaken, the authority must take into consideration what is reasonable having regard to the likely costs involved, the seriousness of the harm and/or pollution in question and the guidance.

Stage VIII – serving a remediation notice. Unless one of the restrictions as outlined in Stage VII applies, the authority is under a duty to issue a remediation notice. An appeal can be lodged within 21 days of service, and fines are in place in respect to non-compliance. Enforcing authorities must keep registers of the remediation notices issued, sites designated as "special sites", appeals notifications etc.

2.4 **UK environmental and regulatory authorities**

Within the United Kingdom, the various environmental legislation and regulations are implemented by a number of local and national Government departments and semi-independent national agencies. The following details the main responsibilities of these authorities, and provides the reader with an insight of the information available from these bodies.

Central government departments

Department of the Environment, Transport and the Regions (DETR)

The DETR is the key central government department with responsibility for environmental matters. Responsibilities include planning, environmental protection, water, countryside conservation, local government, housing and construction. It is therefore clear that the DETR is not simply concerned with environmental protection, and it has been argued that because of its wide remit of responsibility other factors tend to take priority over environmental protection.

The Secretary of State for the Environment has wide legislative and quasi-legislative powers, which stem from the framework nature of the main environmental protection legislation and from the need to update legislation in light of EU requirements.

The DETR may impose its policies in a number of ways other than directly changing the law. These include making directions, the power to approve actions of regulatory bodies, or the power to hear appeals. Alternatively, since the DETR and the Treasury have complete control for the budgets of a number of regulatory agencies (*e.g.* the Countryside Commission), they can manipulate the available resources.

Department of Trade and Industry (DTI)

The DTI is mainly responsible for policy in relation to industry, business, energy and trade, although it also has a wide remit with regard to the environment, with relevant divisions including the Environment and Energy Technologies Division and the International Trade Policy Division. Many of the policy decisions of central importance to global warming and acid rain are made by the DTI. The Warren Spring Laboratory, which has specialist expertise in pollution abatement technology, merged with AEA Technology in 1994 to form the new National Environmental Technology Centre, which in turn operates the "Environmental Helpline", a free telephone enquiry service set up by the DTI for companies wishing to find out more about environmental issues that may affect their business.

Ministry of Agriculture, Fisheries and Food (MAFF)

The MAFF is responsible for some important environmental controls and policy matters through its Countryside, Marine Environment and Fisheries Directorates. The MAFF not only controls dumping and incineration at sea via the use of licensing, along with other marine environmental protection measures, but also deals with the effects of pollution on agriculture, conservation policy and the environmental implications of agriculture and horticulture. The MAFF has issued Codes of Good Agricultural Practice which relate to the pollution of water, air and soil, along with regulations setting construction standards for waste storage facilities on farms.

National bodies

There are a number of national bodies that have been set up by the Government, and in most cases are staffed by the civil service.

Correspondingly, these bodies cannot be classed as non-governmental organisations.

Biotechnology and Biological Sciences Research Council

This is a non-departmental public body which was established by Royal Charter in 1994 by the amalgamation of the former Agricultural and Food Research Council and the biotechnology and biological sciences programme of the former Science and Engineering Research Council. The body is principally involved in promoting and funding research, providing advice and promoting public understanding.

Countryside Commission

The Commission has responsibility for the conservation of the English countryside. It is involved in the designation and management of Areas of Outstanding Natural Beauty and National Parks and must be consulted when a development requires environmental assessment or otherwise affects a designated area.

Joint Nature Conservation Committee

The Committee was established under EPA 1990 to co-ordinate nature conservation in the United Kingdom. It serves as a forum for the three country nature conservation agencies (English Nature, the Countryside Council for Wales and Scottish Natural Heritage) and it is through this forum that the agencies carry out their statutory duties (special functions) for the United Kingdom.

Countryside Council of Wales

The Council is responsible for the conservation and enhancement of the Welsh countryside, combined with a regard for the socio-economic interest of rural areas. The Council also advises the Government on matters relating to wildlife, countryside and maritime conservation matters in Wales and is the executive authority for the conservation of habitats and wildlife. In addition, it selects and notifies Sites of Special Scientific Interest and is a statutory consultee when development affects such sites.

English Nature

English Nature (previously the Nature Conservancy Council for

England) promotes the conservation of English wildlife and nature conservation through advising the Government and running National Nature Reserves. In addition, it selects and notifies Sites of Special Scientific Interest and is a statutory consultee when development affects these designated sites.

Scottish Natural Heritage

This body is responsible to the Secretary of State for Scotland and was established as a statutory body in 1991 to take over the combined responsibilities of the Countryside Commission for Scotland and the Nature Conservancy Council for Scotland. Duties undertaken by it include designating and managing sites to protect habitats and wildlife, and promoting conservation and public enjoyment.

Health and Safety Executive (HSE)

The HSE was first set up in 1975 and was responsible for implementing the provisions of the Health and Safety at Work Act 1974. The HSE reports to the Health and Safety Commission, which in turn is responsible to appropriate Ministers for the supervision of the 1974 Act. The Commission also reviews health and safety legislation and submits proposals for new or revised legislation.

Natural Environment Research Council

The Council is responsible for organisating and developing research in the physical and biological sciences which relates to the natural environment. It gives advice to a number of Government departments and carries out research via its own grant-aided institutions, grants to universities and collaboration with other organisations and institutes.

Royal Commission on Environmental Pollution

The Commission advises and reports on environmental policy and has to date produced a number of reports on a wide variety of matters. The reports have enormous authority in relation to the subject-matter discussed and exert a significant influence on the direction of environmental policy.

Local government

In the majority of cases, the implementation and enforcement of environmental legislation falls under the remit of local government. In England there is both a two-tier system incorporating county and district or borough councils and a one-tier system operational, for example, in metropolitan areas such as London, Birmingham and Manchester, where the same duties are undertaken by unitary authorities, sometimes described as "borough councils". In National Parks and Designated Development Areas there are authorities specifically responsible for a number of functions usually undertaken by the local authority (*e.g.* planning control). In Scotland, the principal authorities are the regions, which are in turn subdivided into districts. In Wales there are unitary authorities.

Local authorities have environmental health departments which are the first port of call for members of the public who wish to make representations about activities which are believed to be adversely affecting the environment. The environmental health officer's principal role is the protection of health (which correspondingly includes air pollution), dealing with complaints relating to nuisances, the consequences of waste accumulations and all the health and safety aspects of places to which the public has access.

County councils

There are a number of county councils in England, with regional councils and island councils in Scotland. In addition there are unitary authorities in Wales. A county council has the responsibility in terms of pollution control for strategic land use planning and development control relating to mineral extraction.

District councils

The district councils form the second tier (with county councils) in the system, so that each county is divided into districts (four or five on average). The districts operate at the more detailed level of delivery of local services and it is at this level where local environmental health is safeguarded and detailed land use planning undertaken.

Unitary authorities

The move towards a unitary or single-tier structure for local government commenced in the 1980s in the metropolitan areas so that, for example, in 1985 the Greater London Council was abolished and replaced by 32 London Borough Councils, or unitary authorities. This has since been extended to other parts of the country. A unitary authority has sole responsibility for land use planning and environmental health, although in some cases a hybrid arrangement exists with the former county council.

Non-governmental organisations (NGOs)

There are a number of NGOs concerned with or involved in environmental concerns. These include, but are not limited to, the following

Business in the Environment

Launched in 1989, the aim of Business in the Environment is to devise practical steps towards applying the principles of sustainable development by encouraging businesses to raise the priority of the environment within the commercial world.

Chemical Industries Association

This is a trade association for the chemical industry, which is normally highly sensitive to changes in environmental legislation. It is active in a number of the working groups on the environment which have industrial representation.

Environment Council

This is a charitable organisation the main activity of which centres around its Business and the Environment Programme. This programme aims to promote the ethos that environmental sense is commercial sense. The Council also runs an Information Programme, which gives guidance on sources of environmental information.

Environmental Industries Council

An independent trade organisation set up in 1995, the Commission lobbies the Government to promote and support UK suppliers of environmental technologies and services.

Friends of the Earth

The main aim of Friends of the Earth is the protection of the environment, the promotion of sustainable alternatives and the dissemination of information to the public.

Green Alliance

The Green Alliance is principally involved in environmental policy and politics. It runs four programmes: industry and finance, networking, politics, and media.

Greenpeace

Greenpeace is an international pressure group which undertakes campaigns against abuse of the environment. Pressure is brought on governments and companies through non-violent actions and protests combined with lobbying.

Industry Council for Packaging and the Environment

Representing all sectors of the UK economy involved in the manufacture and packaging of goods, the Council aims to enhance the protection of the environment while at the same time protecting members from unjustified attacks on environmentally sensitive issues.

Institute of European Environmental Policy

The Institute is principally a research body which works closely with the European Commission to develop social innovations and improve the quality of life.

Marine Conservation Society

The Society aims to protect the marine environment and promote its practical management through lobbying and campaigning on marine issues. It also conducts research on marine life and supervises the safeguard of a number of coastal sites.

National Society for Clean Air and Environmental Protection

This body is principally aimed at improving the environment by promoting clean air, noise reduction and implementing other pollution control measures. It combines expertise from government, technical and academic sources.

The National Trust

The Trust is an independent charity which preserves and protects places of historic interest or natural beauty for England, Wales and Northern Ireland.

Environmental agencies

Environment Agency (England and Wales)

In October 1991, the Government issued a consultation document, *Improving Environmental Quality*, which proposed the establishment of an environmental agency which had control over air, land and water pollution. Correspondingly, in December 1994, the Government introduced the Environment Bill primarily to set up the new Environment Agency for England and Wales, with a separate Environment Protection Agency for Scotland.

Part I of the Environment Act 1995 establishes the Environment Agency for England and Wales and transfers the functions, property, powers, rights and liabilities of HMIP, the NRA, certain smaller units of the former DoE and the local Waste Regulation Authorities, together with certain property, rights, functions and liabilities of the Secretary of State for the Environment, to this body. Note that section 2(3) of the Act actually abolished the NRA. Local authorities remain responsible for those air pollution controls for prescribed processes previously not dealt with by HMIP.

The Agency is responsible for the protection and improvement of the environment in an integrated manner and has major responsibilities for the control of industrial pollution and wastes and for the regulation and enhancement of the water environment. In addition to these responsibilities, the Agency has an enhanced remit under the 1995 Act including the National Waste Strategy, Producer Responsibility for Waste and the production of integrated "State of the Environment" reports.

The 1995 White Paper and Annual Report *This Common Inheritance* suggested that the Agency take forward the Government's policy of sustainable development in three important ways:

(a) by reducing the number of regulatory authorities that industry has to deal with to provide a "one stop shop" for environmental regulation;
(b) by strengthening integrated approaches to pollution control; and
(c) by becoming respected centres of expertise and excellence on environmental monitoring, research and science.

The underlying principle for the Agency's development is for organisations which previously had to deal with a number of regulatory bodies to deal only with a single point of contact, and receive through customer service centres an integrated co-ordinated service. The Environment Agency's main functions and responsibilities include:

(a) implementing and controlling industrial processes governed under IPC, as set out in Part I of EPA 1990 (to be replaced by IPPC);
(b) maintaining public registers (*i.e.* IPC, industrial works, waste management licences, carriers and brokers of controlled wastes, water quality and pollution control, water abstraction and radioactive substances);
(c) providing advice to the Environment Secretary relating to the National Air strategy and providing guidance to local authorities concerning Air Quality Management Plans;
(d) managing and regulating the remediation undertaken on contaminated land identified as special sites;
(e) regulating and controlling waste management sites and carriers;
(f) implementing the Government's National Waste Management Strategy;
(g) developing the Government's radioactive waste management strategy;
(h) conserving and managing the appropriate use of water resources, including granting abstraction licences;
(i) providing and monitoring discharge consents to water (surface waters, including coastal waters and groundwater);
(j) conserving the water environment, including Areas of Outstanding Natural Beauty or of environmental sensitivity, and promoting its use for recreation;
(k) managing fisheries resources, including issuing angling licences; and
(l) overseeing all flood defence concerns.

Scottish Environment Protection Agency

This Agency undertakes a similar function to that of its counterpart in

England and Wales. The Scottish Agency came into effect on 1 April 1996 and combined the roles of HM Industrial Pollution Inspectorate, the River Purification Authorities, some duties of the Hazardous Waste Inspectorate and the waste regulation and local air pollution control functions of district and island councils.

European Environment Agency

The European Environment Agency, unlike the UK agencies, is not a law enforcement agency, but rather was established to co-ordinate an information network between Member States and to provide them with the objective information necessary to formulate and implement effective environmental policies. Areas of work which have been targeted for emphasis so that effective environmental policies can be developed are:

- Air and atmospheric emissions
- Water quality, pollutants and water resources
- Soil, flora and fauna
- Land use and natural resources
- Waste management
- Noise emissions
- Chemical substances which are hazardous to the environment
- Coastal protection.

Chapter 3

· The Audit Process ·

3.1 **Introduction**

There are a number of different environmental auditing procedures advocated in auditing literature. Many of these proposed procedures contain similar elements and follow similar structures. This chapter sets out to describe a general environmental auditing procedure, derived from a synthesis of the available procedures. In doing this, it draws heavily on the most accepted model of the audit procedure first developed by Arthur D Little and then later adopted by the ICC. In addition, this section places specific emphasis on the due diligence auditing methodology, so that the reader can apply directly the relevant sections to any future audit to be undertaken.

The environmental auditing process can be divided into three main areas of activity:

- Pre-audit activities
- On-site audit activities
- Post-audit activities.

3.2 **Pre-audit activities**

Once a commitment to auditing has been undertaken there are a number of activities that need to be completed before the audit proper can commence. These activities are aimed at reducing the amount of time spent at any particular site, as the time spent on a site is costly to both the auditors and the auditees.

Selecting the facilities to be audited

If the audit is part of an overall audit programme, the sites that are to be audited need to be determined, the order being dictated by corporate preference and logistical issues of geographical proximity.

Scheduling and informing facility management of decision to audit

The facility's management should be made aware of the date of the audit as soon as possible, enabling them to adjust and become used to the concept. However, in some companies little or no warning is given of an imminent audit. Companies employing such tactics feel that using a surprise audit enables a true evaluation of the facility's operations as they routinely function. Furthermore, care must be taken that only appropriate personnel are informed of the audit, as the true reason for the audit being undertaken, especially within the area of due diligence auditing, may not be known by the majority of personnel. An example of this scenario would be a company take-over, merger or divestment.

Identifying audit scope

The audit scope needs to be established, as it identifies the extent and limits of the audit in terms of factors such as physical location and organisational activity, as well as the manner of reporting. The auditee should usually be consulted when establishing the scope and any subsequent changes to it would require the agreement of the client and/or the management responsible for the facility being audited.

Planning the audit

The audit plan should be designed to be flexible in order to permit changes in emphasis based on information gathered during the audit and to permit effective use of resources. The plan should include (amended from ISO 14011:1996):

- The audit objectives and scope
- The audit criteria
- Identification of the auditee's organisational and functional units to be audited
- Identification of the functions and/or individuals within the auditee's organisation having significant direct responsibilities regarding the audit scope
- Identification of those elements of the audit scope which are of a high audit priority
- The procedure for auditing those areas identified in the

objectives of the audit as appropriate to the auditee's organisation
- The working and reporting language of the audit
- Identification of reference documents
- The expected time and duration for the major audit activities
- The dates and places where the audit is to be conducted
- Identification of audit team members
- The schedule of meetings to be held with the auditee's management
- Confidentiality requirements
- Report content and format, expected date of issue and distribution of the audit report
- Documentation retention requirements.

All parties involved in the auditing process should be informed of the audit plan, including the auditee, the client and the audit team members, with approval of the audit plan being sought from the client. If the auditee objects to any element of the audit plan, then the objections should be made known to the lead auditor, and resolved between the lead auditor, the client and the auditee.

Selection of audit team and assignments

The audit team leader and audit members need to be selected and their availability established. As appropriate, each audit team member should be assigned a specific element of the audit and be instructed on the audit procedure to follow. Such instructions should be made by the lead auditor in consultation with the audit team member concerned.

Collection of chosen working papers

The working papers required to facilitate the auditors' investigations may include, but are not limited to:

(a) forms for documenting and supporting audit evidence and findings;
(b) procedures and checklists used to evaluate the audit findings; and
(c) records of meetings.

Collecting background information on the facility

This involves obtaining as much information as possible about the facility's organisation, layout and processes, the relevant regulations and standards that govern its operation and the company policies that cover it. This information can be obtained by either undertaking an advance visit to the facility and/or by circulating a pre-survey questionnaire to its personnel. The main purpose of an advance visit is that it enables the auditors to develop a basic understanding of the facility's processes and its environmental management system, and to brief the facility staff on the objectives of the upcoming audit.

Collecting background information on site's historical uses, location etc

The site's past uses and activities historically undertaken on it should be established with particular attention being paid to the likelihood of soil/groundwater contamination. In addition, information relating to the geology, hydrogeology and proximity to sensitive targets (i.e. controlled waters, residential development, Sites of Special Scientific Interest) should be obtained. Since the sourcing of this information can be a lengthy process it is advised that enquiries relating to this information be initiated as soon as possible within the auditing process, to ensure that unnecessary delays are not caused through lack of information at later stages in the audit.

3.3 On-site audit activities

Opening meeting and initial tour

The opening meeting and initial tour are especially important if a pre-audit visit to the site is not possible. The opening meeting represents the first meeting between the audit team and the facility personnel. The meeting should be used to help reassure facility personnel about any apprehensions they may have in regard to the audit. During the meeting the individual members of the audit team should be introduced to the auditee's management and the lead auditor should explain the purpose, scope and conduct of the audit, in addition to outlining the activities of each of the individual auditors to the facility personnel. The official communication links between the audit team and the auditee should be established and the resources and facilities needed by the audit team should be confirmed. The pre-survey questionnaire should also be reviewed with the facility staff to clarify any misunderstandings or mistakes which may have been made. Following this, the facility personnel should then present an overview of the facility's organisation and operations, including relevant site safety and emergency procedures.

During this opening meeting it is important that the audit etiquette, which should be followed by the audit team members during on-site activities, is established. The establishment of this protocol should involve the site management detailing any health and safety rules that the audit team should follow, identifying any specific safety equipment that the team may require (*e.g.* fluorescent jackets, hard hats etc) and identifying any "no-go" areas. The team should also identify whether there are any restrictions to taking photographs during the audit, and should request permission to take photographs before on-site investigations begin for two specific reasons:

(1) Some industrial activities are commercially sensitive and for this reason the company may not want photographs to be taken in certain areas of the facility or of certain processes. Such areas should be determined and not be photographed during the audit.
(2) In some industrial installations handling flammable materials a "hot-work" permit is needed for all contractors. The permit controls the use of equipment which might generate a spark, or act as a source of ignition. In hot work areas special cameras can be used which are insulated and cannot generate a point source of ignition.

The opening meeting should, if possible, be followed by a short orientation tour of the facility. The key issues at the facility which the team should pay particular attention to are:

- All points of emissions
- Storage of raw materials or finished products
- Process operations
- Evidence of ground contamination
- Environmental monitoring
- General housekeeping
- Training and information
- Record-keeping

- Wastes management
- History of spills
- Leaks and accidents
- Administrative offices
- Other places of importance.

Understanding the management system

This step of the on-site phase requires the audit team to develop a working understanding of how the facility manages the activities that influence the environment and how any Environmental Management System, if there is one, works. To do this an auditor must understand the facility's procedures and processes; its internal environment management and engineering controls; the organisation of facility staff and their designated responsibilities; any compliance parameters; and any past or present environmental problems encountered at the facility (ICC, 1991).

Such an understanding can be obtained from the information gathered in the pre-audit stage, through discussions and interviews with facility staff, and through a limited amount of verification testing. At the end of this stage the auditors should be able to record and demonstrate their understanding of the facility's management system by compiling a flow chart and/or writing a narrative description of the system. The flow chart/description should be documented in the working papers.

The development of an understanding of the internal management system is crucial to the audit as it provides a basis against which an auditor can assess the strengths and weaknesses of the company's internal controls, and develop appropriate schemes for testing the effectiveness of these controls.

Assessing strengths and weaknesses of internal controls

At this stage in the audit process, the strengths and weaknesses of the facility's internal management controls, and the risks associated with their failure need to be established. This is done to determine whether the internal controls are adequate to achieve the desired level of environmental protection and compliance. Internal controls consist of both the management procedures and the equipment or engineered restraints that influence the facility's environmental performance.

There are seven different types of indicator that can be used by an auditor to determine the effectiveness and suitability of internal controls. The use of these indicators requires considerable judgement on the auditor's behalf, as there are no quantifiable standards against which they can be compared. A company's internal controls will be acceptable if:

- Facility personnel are well trained and experienced
- Responsibilities are clearly defined and carefully assigned
- Duties are divided to minimise conflicts of interest and to build in checks and balances
- Authorisation systems are in place
- Internal verification procedures are established
- Protective measures are taken (*e.g.* security systems and alarms installed)
- Procedures and compliance/exemption results are documented.

The aim of this phase is to determine upon which specific aspects of the facility the auditors should focus their attention. Areas that are shown to have high environmental risk and which are governed by weak internal controls deserve intense examination, while those areas posing a small environmental risk and which have strong internal controls need relatively little investigation. This part of the audit drives the subsequent gathering of audit evidence.

Gathering audit evidence

Compliance of a facility with legislation, regulations or company policy can only be determined on the basis of evidence collected at this stage of the audit process. Audit evidence is collected in a variety of ways including reviewing records, examining available documents and logs, conducting emissions sampling and interviewing staff. The collection of this evidence provides the means by which any suspected weakness identified in the previous stage of the on-site audit are confirmed or disproved. The ICC (1991) identifies three main procedures for collecting audit information:

(1) *Enquiry*. The auditor can question staff on either a formal or informal basis. Audit protocols are often used to help achieve this. Enquiry presents a relatively easy means by which to obtain information, involves minimal cost, is relatively quick and

straightforward to undertake and provides feedback quickly. However, this means of obtaining information does not by itself provide definite proof of compliance.

(2) *Observation*. Information about the facility can be obtained from the auditors' senses of sight, smell and hearing. This form of physical examination is frequently undertaken as it provides an extremely reliable source of information.

(3) *Verification testing*. This considers a broad range of activities from retracing and re-computing data to checking the calibration of monitoring and sampling equipment. This is the most sophisticated method of obtaining evidence, but it is also time and resource intensive. Verification testing involves the auditor undertaking systems checks on either the management system or the control equipment. By actually testing the systems the auditor is able to determine whether what should happen in theory does actually happen.

A camera is an extremely useful tool in gathering audit evidence, as it can be used to graphically illustrate a problem at a site or a situation of non-compliance. An auditor should always carry a camera and should use it freely, as the pictures taken often form a useful means to jog the memory once the auditor is off-site and compiling the final report. Alternatively the site inspection may be recorded on camcorders or digital camera and the tapes/disks filed for later reference.

Communication during the audit

One of the principle methods by which an auditor can gather audit evidence is via enquiry. It is therefore important for an auditor to have effective interpersonal skills, so that communication and relationships with people on the facility allow the most information to be obtained.

One of the principle objectives of an audit is to minimise the time spent on site, and the associated minimal disruption to on-site operations requires a co-operative spirit to be established between the auditor and the facility personnel. Although there may be explicit directives from upper management that personnel are to be as co-operative as possible, it is more important that a personal relationship is established between the audit team members and the facility personnel. Each auditor should therefore seek to establish a positive working relationship with the site contacts. The ICC describes four key characteristics which it believes a good auditor should possess:

(1) *Candour*. The auditor should tell the auditee what to expect from the audit and then be fully candid in communicating all impressions – positive or negative. Tensions created by a feeling on either side that full communication is not taking place can severely hamper the audit.

(2) *Communication*. The auditor must be a good listener, exhibiting a genuine interest in both the person contacted and the communication taking place. The auditor should be alert for key information and should keep the interview directed toward efficiently accomplishing its objective.

(3) *Attitude*. A friendly, professional attitude on the part of the auditor is much more likely to elicit co-operation than an overly aggressive, autocratic one. The auditor must be firm and persistent in gathering all needed information, although the manners of a good guest serve the auditor well.

(4) *Empathy*. The auditor should keep in mind that the audit can be a stressful event to most auditees. He should work to set the persons contacted at ease. If the auditor demonstrates that he understands that the auditee has difficulties to overcome in performing his job and even in setting aside the time for the audit interview, barriers to communication will be lowered.

Evaluation of audit evidence

This stage involves the evaluation of audit evidence to identify any significant observations or exceptions to the compliance parameters. First, the individual auditors must review their compiled evidence and their associated protocols to determine whether sufficient evidence has been collected for the purpose of the audit. After this has been done, and any gaps in the evidence filled, the auditor must then evaluate the evidence and compile a list of exceptions and observations for each of the topic areas studied. This list must be discussed among the team as a whole, and an integrated list of all of the auditors' findings must be compiled. By integrating the list, larger trends or patterns in audit findings may be revealed. All audit findings must be substantiated by the evidence, and care must be taken to ensure this.

Close out/exit meeting

Once all of the findings have been determined by the audit team, an exit meeting should be held with the facility personnel. The purpose of this meeting is to communicate the results of the audit findings, and to help resolve any misunderstandings, misinterpretations or errors that the auditors may have made. With the completion of this meeting the on-site auditing activities are finished.

3.4 Post-audit activities

Post-audit activities begin with the preparation of a draft report. The draft report can actually be prepared on the site itself, but most usually it is written immediately after the on-site audit. The nature of this report is described in Chapter 7. Once the draft report has been produced, it should be reviewed by the facility personnel directly involved in the audit (*e.g.* the site management), and from this a final report should be derived. This final report should then be distributed to all interested parties within the company. It is often useful to establish a circulation list at the outset (*i.e.* when commissioned to undertake the audit) so that on-site staff are aware of the parties who will have access to the information provided by the audit.

Once the final report has been issued the formal role of the audit team is completed. However, the audit process continues as there is still a need to act upon the report and its associated recommendations. This part of the process is essential, as "Audit programmes should not be undertaken unless there is a prior commitment to address all of the findings" (ICC, 1991). The benefits of an audit would be lost if no action was taken to follow up the report and implement its recommendations. An action plan should be prepared by management to implement the audit findings. This action plan provides a means of orchestrating any actions that need to be undertaken, and should assign responsibilities for undertaking corrective action, determine potential solutions to the problems posed and establish a time scale over which actions should be completed. In some cases the action plan can be integrated into the final report.

It is important that any action plan is followed up and monitored by either the auditors, the environmental staff or the operating management. The ICC (1991) identifies five elements of a successful follow-up programme:

(1) A standard action plan format.
(2) Established procedures for approving the action plan and communicating its contents.

(3) Regular reporting of the action plan's status (*i.e.* showing what action has been taken and what action is imminent).
(4) Special reporting and chasing up of overdue action.
(5) Independent auditing of the action plan to verify that all actions sanctioned have been completed.

When the action plan has been completed, the formal audit procedure is complete.

Chapter 4

Collecting Background Information

4.1 Determining a site's past and present uses

Given the need to determine the presence of potentially contaminative uses it is important to collect as much information about the history of a site as possible.

There are two main sources of information relating to a site's history which can readily be accessed: historical maps and planning records. Note that there are other means, but they are less accessible and more prone to variation than the maps and records. These additional means are described in detail in the DoE's Contaminated Land Research Report No 3, *Documentary Research on Industrial Sites* (1994).

4.2 Historical maps

Historical maps are an invaluable and recognised source of information about a site. Despite possible variations in accuracy owing to survey methods, printing and scale, they provide one of the best guides to determining the past uses of a site, and thereby in identifying potentially contaminative uses. The prime source of historical maps for a site are from the Ordnance Survey, a body that was established in 1791.

Between 1840 and 1945 England and Wales was mapped on a county basis at a scale of 6" and/or 25" to the mile, and all counties in England and Wales were surveyed at least twice during this period. The first series of maps, referred to as the "County First Series", were completed on a national basis by 1893 (work started on this series in 1840). It was originally intended to revise these maps on a 20-year cycle, and a first revision was completed by 1914 (the County Second Series). Further revisions were also begun in 1904 and 1911 (the County Third and Fourth Series), but these were never completed for the whole of England and Wales.

From 1946 onwards mapping was undertaken (at a variety of scales) on a national basis, and a national grid was introduced. Since this time revision of maps has been on a continuous basis, with new maps published as and when required. Urban areas tend to have been surveyed more frequently, with most areas having being mapped five times up to the present day, since development in urban areas (in general) is more intense than the rural areas.

Given the above, there tends to be good map coverage of most sites in the United Kingdom from 1911 onwards.

Historical maps are available at a number of places. The best sources, however, tend to be either the local reference library (which often has restricted coverage prior to 1945) or the county archive (which usually has excellent coverage pre-1945). The Map Room at the British Library is also an excellent source of maps. There are several other major national collections of maps, such as those held by the Bodlean Library (Oxford) and the Cambridge University Library. More detailed information about maps and their interpretation can be obtained from Contaminated Land Research Report No 3 (1994).

4.3 **Planning records**

Detailed information about a site can often be obtained from local authority planning registers, which contain the details of all planning consents issued for a site under the various planning legislation. Planning consents are usually available for a site from 1947 onwards, the time at which the Town and Country Planning Act 1990 (the main body of planning law) was first enacted. Planning records and consents can provide specific information on a site, such as the details of process plant used on-site, the nature of operations undertaken on the site, storage facilities present on-site (*e.g.* petroleum tanks) and the materials used in construction (asbestos in particular). Such information can give the auditor an insight into the potential nature of any contaminative liabilities associated with a site, and may also add detail to information obtained in the historical search.

Example 1

Brownfield Developments Ltd are looking to acquire a site for warehousing. They have identified a site in Cheshire which is presently developed as a warehousing unit and suits their location and distribution requirements. The auditor is required to undertake a due diligence Phase I environmental audit on the site. On visiting the site, the auditor notes that it is well run, with no evident operational liabilities associated with the site. Upon undertaking investigations into the historical development of the site, he visits the local reference library, which holds detailed historical maps for the area. The historical maps set out in Table 4.1 are available for his inspection. From these maps the historical development of the site can be assessed in part.

Year	Sheet	Scale	Figure
1891	Lancashire First Edition Sheet CXV.9	1:2,500	4.1
1908	Edition of 1908 Sheet CXV.SW	1:10,000	4.2
1926	Enlargement of 1926 Revision Sheet CXV.9.NE	1:1,250	4.3
1937	Revision of 1937 Sheet CXV.9	1:2,500	4.4
1958	Plan SJ 58 NW	1:10,000	4.5
1969	Sheet SJ 5185	1:2,500	4.6
1995	Sheet SJ 2285 SW	1:1,250	4.7

Table 4.1: Historical maps for Example 1

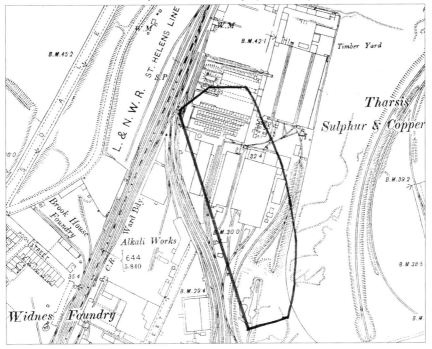

Figure 4.1

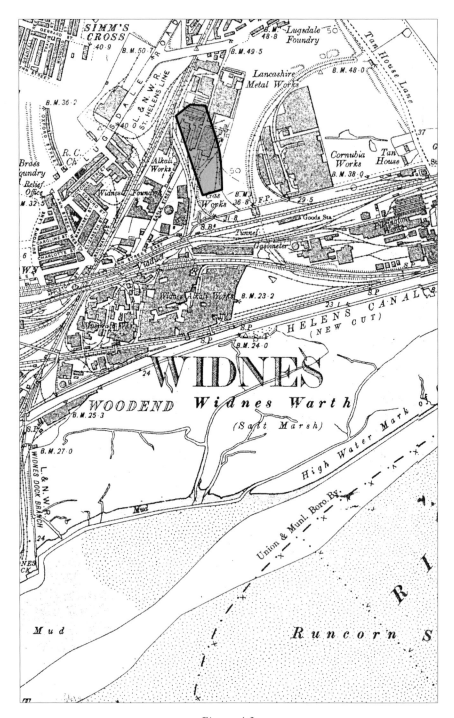

Figure 4.2

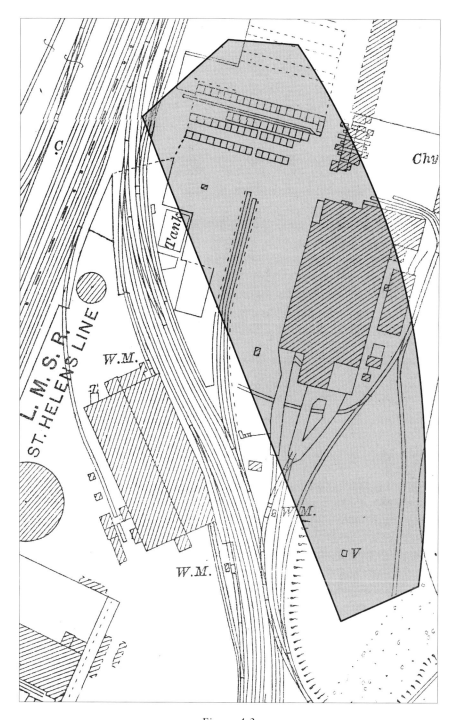

Figure 4.3

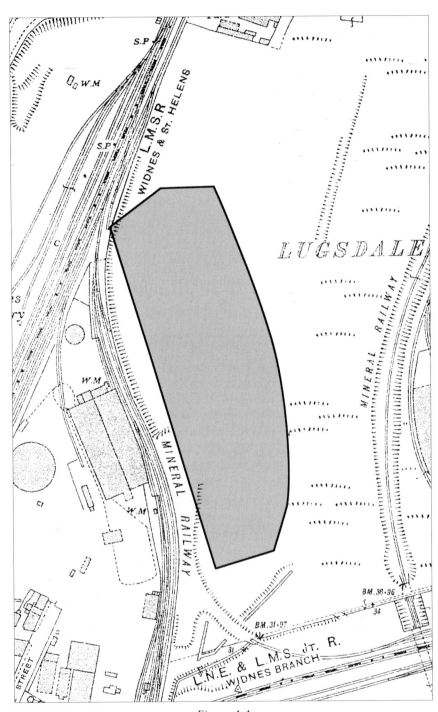

Figure 4.4

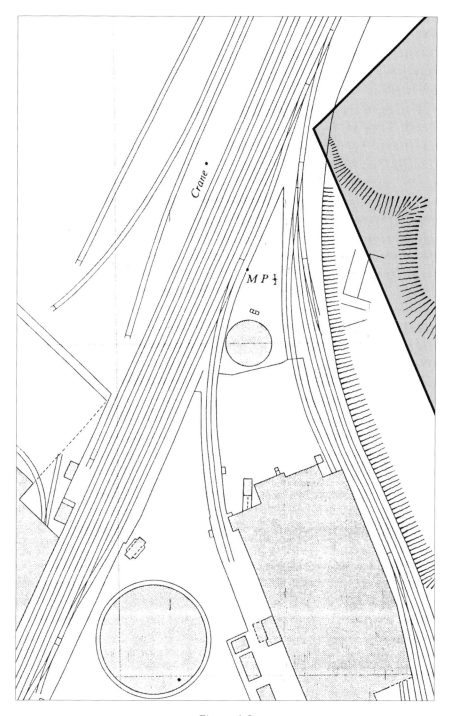

Figure 4.5

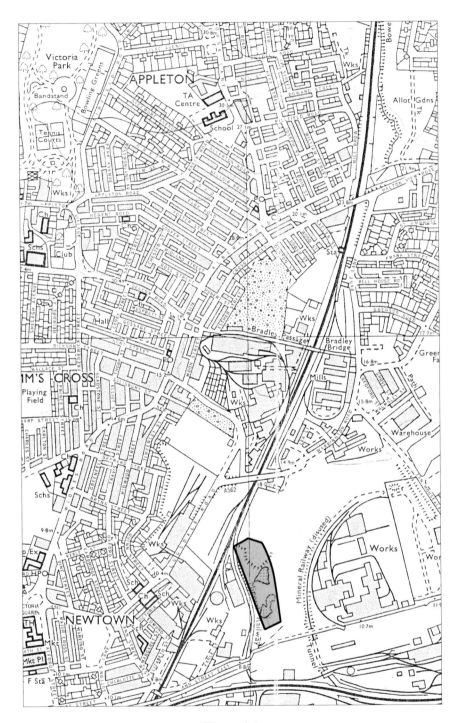

Figure 4.6

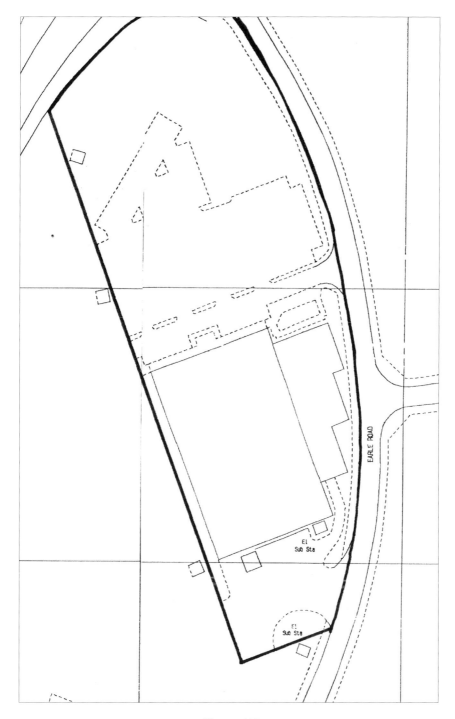

Figure 4.7

As can be seen from the 1891 map (see Fig 4.1), the site is located predominantly upon the Tharsis Sulphur and Copper Works. The site is occupied by a number of buildings and is also crossed by a number of railway lines. The southern part of the site appears to be partially excavated, and the rail lines in this area are elevated with a small artificial pond also noted.

The 1908 map (Fig 4.2) shows that the site has undergone a change in use and is now identified as the "Lancashire Metal Works". A number of rail lines still traverse the site. The area surrounding the site is shown to be heavily industrialised with a sulphate of copper works (the "Cornubia Works"), a gas works, a foundry and an alkali works all within close proximity of the site's boundaries.

Although the 1926 map (Fig 4.3) shows that the site is still occupied by the Lancashire Metal Works, the site appears to be disused. The surrounding area remains highly industrialised.

As can be seen from the 1937 map (see Fig 4.4), the site is shown to have been cleared and is unoccupied, with no marked use. The surrounding area remains principally developed for industrial usage.

The 1958 map (Fig 4.5) only shows the north-westerly tip of the site. From this map, however, it can be determined that the site is unoccupied, although it can be seen that there seems to be a scarp face on the part of the site shown. This would seem to indicate either the infill of the land or the extraction of material from the site.

The 1969 map (Fig 4.6) shows the site in its entirety. As can be seen from this map a scarp slope is still shown to be present on the site. Also present are two small ponds. The surrounding area remains heavily industrialised.

The 1995 map (Fig 4.7) shows the site as it is now.

Planning details for the site are also available from the local council offices. Planning applications for the site are reviewed, the details of which are summarised in Table 4.2. In the planning application dated 1987 there was a letter from the National Coal Board to the planning authority. This letter stated that there was a seam of coal at considerable depth beneath the site which was once worked in 1988. The Board stated that any subsidence from this working should have now ceased (the letter was dated 11 January 1988). The letter then goes on to note that there are other seams beneath the site which are designated as to be worked in the future.

Application date	Details of application
2/87	Erection of retail warehouse, garden centre and new access
2/87	Outline for retail warehouse, food store, garden centre and parking
12/87	New food retail warehouses, garden centre, car wash, parking, landscaping and new accesses
2/89	Wall mounted signs illuminated by external lighting
3/94	Free standing advertisement board
11/96	Replacement fascia signs – in this application reference is made to the sites use as a former "pre-76 waste disposal site" and its use as a chemical waste heap

Note: For confidentiality reasons we have not included details relating to the application **number and applicant**. Ordinarily, this information would also be included.

Table 4.2: Relevant planning applications for Example 1

Summary of site history

From the available information relating to the site it would appear that the site has undergone a number of uses in the past. The site's earliest recorded use was in 1891 when the site was shown to form the main part of the Tharsis Sulphur and Copper Works. The site is then shown to change use and is shown by 1908 to be part of the Lancashire Metal Works. By 1928 the Metal Works are noted to be disused, and by 1936 the site is known to be clear of all buildings. The site remained in this state until approximately 1988 when the site was developed for its current use. During the period 1936 to 1988, there would appear to be have been some activity on the site, as scarp slopes and ponds are shown on the interim map editions. At some point after 1988 the present development was built and the site was used as a warehouse facility.

Virtually all of the historical maps show the surrounding area to be highly industrialised, and it is known that the site has been surrounded by gas works, iron and steel foundries, alkali works and copper sulphate works in the past. On the basis of this information, it is evident that there is a potential for the site to be contaminated.

Example 2

In this instance, the auditor has been commissioned by a housing association to review the historical development of a site which is presently clear, and which is scheduled for redevelopment for residential purposes. The investigations are being undertaken as a condition on the planning application, which states that "prior to any development works, the site shall be subjected to a detailed scheme for the investigation and recording of contamination and a report shall be submitted to and approved by the local Planning Authority", which has been granted subject to the planning conditions being satisfied. Since the site is presently clear of buildings, the auditor has been commissioned to undertake an initial desk-top assessment of the site, reviewing its historical development and consulting the regulatory bodies. Consequently, he visits the British Library and collects the historical maps set out in Table 4.3.

Year	Sheet	Scale	Figure
1870	London sheet LXIV	1:2,500	4.8
1894	London Sheet CV	1:2,500	4.9
1916	London Sheet XS	1:2,500	4.10
1937	London Sheet XII.31	1:2,500	4.11
1970	Plan TQ3776 NE	1:2,500	4.12

Table 4.3: Historical Maps for Example 2

The 1870 map (see Fig 4.8) shows the site and the surrounding area to be highly developed. The site of interest can clearly be seen to be part of an engine works. Close examination of the map shows there was once a "tank" present in the middle of the eastern edge of the site. The Kent Water Works can be seen approximately 60 metres to the west of the site, and dense residential development surrounds the site to the north and the south.

The 1894 map (Fig 4.9) shows that the layout of the site has not changed since the previous map. However the site is clearly referred to as a Marine Engineering Works. No reference is made to the tank that was shown in the previous map. Few changes are apparent to the composition of the surrounding area, although note that a number of the reservoirs belonging to the Kent Water Works appear to have been infilled.

The 1916 map (Fig 4.10) again shows no significant changes in terms of the layout of buildings on the site. However, no reference is made as to the nature of the site's use. The surrounding area shows no significant changes, although the water works to the south now appear to be part of the Metropolitan Water Board.

The 1937 map (Fig 4.11) shows the site to have undergone change, and more buildings are present on the area of interest. As can be clearly seen the site now forms part of a Tin Box Works. The surrounding area shows no significant changes that may affect the site.

The next available map for the site is from 1970 (see Fig 4.12). The site has changed use again, and now forms part of the Thames Iron Works. As can be seen from the map a tank is shown to be present at the southern part of the site. An electricity sub-station is also shown to be present on the centre of the eastern edge of the site. In addition to the iron works which continue into the land to the west of the site, a printing works is shown to be present immediately to the north of the site and a bakery to the east.

A visit is then made to the local authority planning department. The records shown in Table 4.4 are available for this specific part of the site. From the available information, it is clear that the site prior to its demolition had been the subject of a number of potentially contaminative uses. Given the nature and sensitivity of the proposed redevelopment, further investigation is required before a satisfactory report can be issued to the planning authority and the planning condition discharged.

Date	Description of application	Cited use
18/12/67	Covered area for loading and unloading	Tin works
21/5/68	Erection of tin plate store and an office buildings	Tin works
14/1/74	Erection of a temporary portable building to be used as a temporary office whilst a new office block is being erected	Iron works
23/3/82	Rebuilding of buildings damaged by fire	Iron works
7/2/89	Demolition of existing buildings on site	Iron works

Table 4.4: Planning applications for Example 2

4.4 New information sources – environmental data providers

Owing to the new Government guidance on contaminated land, and the now accepted requirement within the due diligence process to assess the potential environmental liability associated with a site, there are now a number of commercial organisations which have identified key

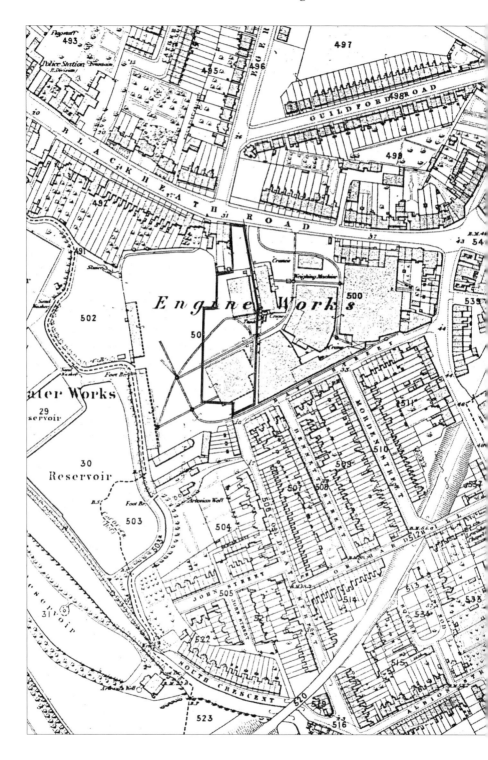

FIGURE 4.8
EXTRACT FROM O.S. MAP
LONDON SHEET LXIV, 1870

SCALE: 1:2500

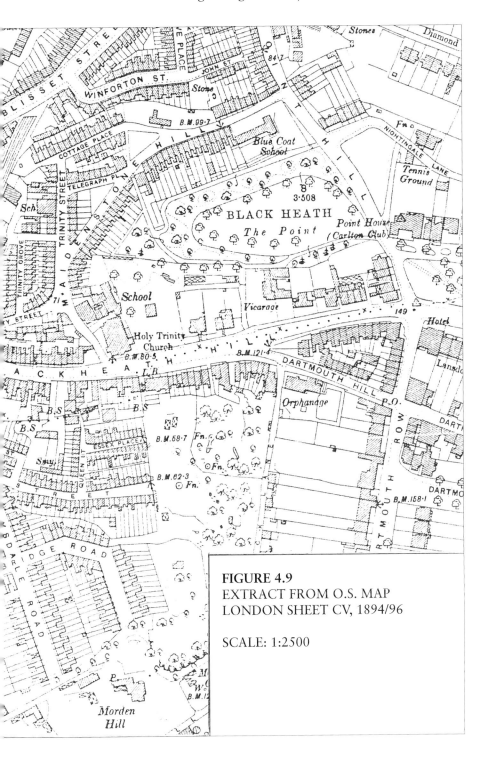

FIGURE 4.9
EXTRACT FROM O.S. MAP
LONDON SHEET CV, 1894/96

SCALE: 1:2500

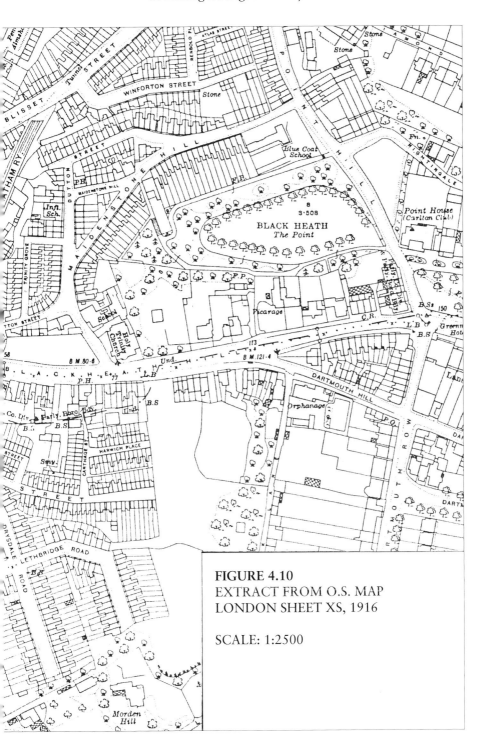

FIGURE 4.10
EXTRACT FROM O.S. MAP
LONDON SHEET XS, 1916

SCALE: 1:2500

FIGURE 4.11
EXTRACT FROM O.S. MAP
LONDON SHEET XII.31, 1937

SCALE: 1:2500

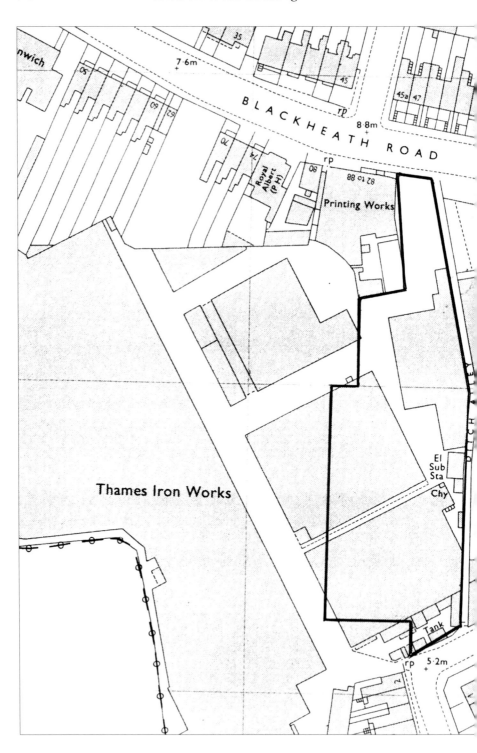

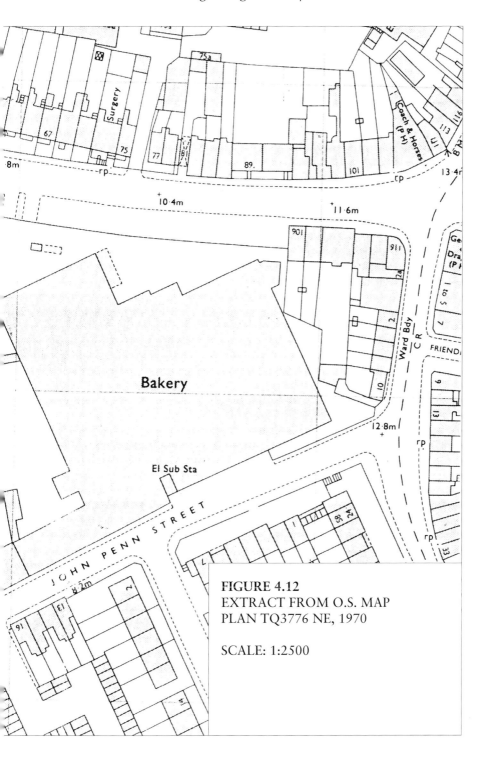

FIGURE 4.12
EXTRACT FROM O.S. MAP
PLAN TQ3776 NE, 1970

SCALE: 1:2500

requirements within the environmental sector and offer a range of services to environmental professionals, all of which can be found in the numerous environmental journals, magazines and periodicals. The range of services include the provision of historical maps and the provision of collected (and sometimes interpreted) regulatory information, which in turn may offer a more cost-effective alternative to having to visit the regional library and/or regulatory bodies. At present the provision of these services is self-regulated by the Environmental Data Association (EDA). This independent body was set up by the individual data suppliers in an attempt to protect the consumer and guarantee both the quality of data supplied and as a minimum the nature of the information required to make an informed judgement. For more information on the individual service providers who meet the EDA's data standards contact the EDA (see "Useful Contact Numbers and Information").

There are a growing number of companies in the United Kingdom who are providing low cost environmental data for use by consultancies, solicitors, surveyors and the general public as an alternative to seeking regulatory responses. There is currently a limited range of data products which can be used to assist the auditor screen properties for environmental risks or provide core environmental data. These products can be divided between those aimed at the commercial property market and those geared more specifically for residential property purchases. Table 4.5 outlines the products currently available in the United Kingdom. These products use large databases and/or geographical information systems (GIS) to search for environmental data and display relevant information in a written report, often combined with maps.

These reports are constantly changing to reflect demand and the geographical areas covered are rapidly expanding. The choice of report depends on which report content is relevant to the intended use. The British Geological Survey is working on a new digital product that will supersede the current ALGI report. It will include several new data sets and will present information in a simpler and more easily understood format. The product is likely to be Internet based and will give an initial risk-based geo-hazard assessment, which may be followed up with a more detailed report if necessary. Landmark is proposing to include data on coal mining in the autumn of 1999, and considering including crime statistics, education, local services and planning applications. Equifax is about to sign an agreement with the Coal Authority to incorporate a coal mining report to compliment its existing products.

Company	Product	Format	Market
British Geological Survey	ALGI Ground Conditions	Report	Residential
British Geological Survey	The Geological Report	Report	Commercial
British Geological Survey	The Radon Report	Report	Residential
Catalytic Data	Sitescope	CD Rom	Commercial
Environmental Auditors Ltd	ContamiCheck	Report	Commercial
Environmental Auditors Ltd	ContamiCheck Homebuyers Report	Report	Residential
Environmental Auditors Ltd	Contamisearch	Report	Commercial
Landmark Information Group	Home Envirosearch	Report	Commercial
Landmark Information Group	Site Search	Report	Commercial
Landmark Information Group	Sitecheck	Report	Commercial
Landmark Information Group	Envirocheck	Report	Commercial
The Coal Authority	Coal Mining Search	Report	Residential
Equifax	HomeSight Radon Report	Report	Residential
Equifax	HomeSight Subsidence Report	Report	Residential
Equifax	HomeSight Landuse Report	Report	Residential

Table 4.5: Commercial and residential environmental data providers in the UK (Taylor 1999)

EAL, which provides the ContamiSearch and ContamiCheck reports, is hoping to include a large body of proprietary data taken from actual site investigations which will add real data on known areas of contamination throughout the United Kingdom.

Another factor that should be taken into consideration relates to the quality of reports. Auditors should be aware that a report is only as good as the data used to generate it and quality control throughout the data input, interpretation etc. The need to manage and ensure a minimum quality of data is one of the EDA's underlying principles, to which Landmark Information Group and EAL are founding members, in addition to being one of only a handful of value added resellers of Environment Agency data. To ensure the results of any data search are not misrepresented in any subsequent report, the auditor should know where the data was obtained, how it is stored and managed and how accurately the database containing the data can be searched. For example, Landmark has a rigorous quality control system and use a sophisticated vector based GIS (Genamap), linked to an Oracle database. EAL uses a simpler, raster-based system to store and search for land-use information, but this is still able to give accuracy to within 100 metres. Both companies regularly update their data to capture changes in the information, and hence are attempting to provide an alternative to the lengthy regulatory data searches which were the norm in the early to mid-1990s.

The variety of information which may be obtained from such sources is wide, and Table 4.6 sets out those data-sets which are currently available.

4.5 **Other environmental sources**

Prior to the development of nationwide environmental databases, the principle sources of environmental data were from the various regulatory and non-regulatory bodies which, by virtue of their operations, held significant quantities of data which were important to the auditor. The following section gives an overview of additional data sources which can be referenced within the audit process in addition to, or as an alternative to, the aforementioned data suppliers. The use of these data sources is very much linked to the individual time scales of each project, the location of any given site and the accuracy of the information required.

Geology

Geological data is available in the form of published maps at 1:50,000 or 1:10,000 scales from the British Geological Survey at Keyworth, Nottingham, in addition to a number of retail outlets. Some areas where modern maps are not published are covered by older, unpublished maps, which are normally available as colour photographic copies. The British Geological Survey also holds a database of borehole logs which can give very useful information relating to the groundwater quality and underlying deposits at various depths. These can be supplied on a CD ROM in index form, or from one of the BGS-recognised resellers of information, allowing the prospective inquirer to obtain information as to the nearest pertinent boreholes, then order detailed borehole logs directly from the BGS.

	ALGI GROUND CONDITIONS REPORT	BGS RADOON REPORT	HOME ENVIROSEARCH	CONTAMICHECK	HOMESIGHT RADON REPORT	HOMESIGHT CLAY SUBSIDENCE REPORT	HOMESIGHT LANDUSE REPORT
Abstraction licenses				1			
BGS boreholes	1			1			
BGS mineral sites	1			3			
BGS survey of waste sites	1			1			
CIMAH and NIHHS sites				1			
Closed (formerly licensed) waste sites			1	1			1
Coal mining (Coal Authority data)			1				1
Crime statistics			3				
Current industrial use			2				2
Currently licensed waste sites			1	1			1
Discharge consents/red list discharges			1				1
EA/SEPA prosecutions/enforcements			1	3			1
Education			3		1	1	1
Flood risk	1		1				
Groundwater vulnerability	1			1			
Hazardous substance consents			1				1
IPC Part A & B prescribed processes			1	1			
Landslide risk	1					1	
Local services information (shops etc)			3		1	1	1
NRPB Radon surveys	1	1	1	1	1		
Operational landfill sites			1	1			1
Petroleum and fuel sites			2				2
Planning applications (possibly contaminating uses)			3				3
Potentially infilled land			1	1			1
Power lines			1				
Radioactive substance consents?			1	1			
Site history (data from historical mapping[2)]	1[3]		2	1			2
Soil classification and leaching	1			1		1	
Solid geology	1			1		1	
Subsidence risk	1		1			1	

1 = 500m search radius; 2 = 250 m search radius; 3 = Proposed inclusion.

Table 4.6: Data sources available in the main UK

Hydrology

The Environment Agency records river quality classifications, flow rates and chemical analysis for most of the major river and surface water bodies. Some of this information is available on a statutory register of water quality maintained by the Agency, or can be provided through a number of the data providers referred to above.

Hydrogeology

The British Geological Survey publishes maps of the whole country and of the major aquifer areas at scales of 1:100,000. Some hydrogeological data can also be sought from the borehole logs described under "Geology" above.

The Environment Agency and the local water companies also hold their own databases of hydrogeological information, but these are generally not available to the public. The Environment Agency publishes groundwater vulnerability maps for the majority of the country, indicating the vulnerability of groundwater to pollution, as it would be affected by surface strata permeability, depth to bedrock etc. Because of the requirement of these maps to assess soil permeability, limited geological information is also presented as text which can be used to confirm the data presented in other sources. Groundwater quality data is maintained on a public register in a similar manner as surface water data.

Licensed water abstractions

The Environment Agency records all licensed surface water and groundwater abstractions. Radial searches of abstractions relating to a particular site can therefore be commissioned (either directly from the Agency or from one of the value added resellers), and data relating to the license holder, the abstraction source, the licensed rate of abstraction and its purpose can be sought. This in turn allows the assessment of possible off-site liabilities associated with a site to be undertaken, should significant contamination be found, and possible migration off-site highlighted.

Discharge consents

Information relating to discharge consents is held by the Environment Agency on all consents to groundwaters and surface waters. This information includes the site owner, the type of discharge, the permitted quantity and the permitted quality limits. The Agency monitors most discharges for chemical quality and this information may be contained on the public register.

Trade effluent discharges

Records of trade effluent discharge consents are available from the local water company, which will be able to advise whether a site holds a consent.

Additional information may be obtained from the Environment Agency relating to the position of Sites of Special Scientific Interest, as well as from English Nature. The Environment Agency also holds information on surface waters used for recreational purposes.

4.6 **Collecting information about a site**

During the pre-audit stages of the audit process an auditor can collect a great deal of information about the site being audited. For example, it has already been shown how information can be collected on the previous uses of the site (see p 69 above). In general, information about a site tends to be available from two sources:

(1) *Internal information*. This relates to the information which is held at, or can be obtained from, the site to be audited. Prior to visiting the site it is useful to request such information so that it can either be forwarded to the auditor prior to the site visit, or be made immediately available at the start of the on-site activities. Internal information can be obtained by the use of a pre-survey questionnaire.

(2) *External information*. Information about a site is held by a large number of organisations, including environmental data providers and the various regulatory authorities. If information is being sought directly from the regulatory bodies, the information can take some time to obtain, and consequently should be requested prior to the site visit. This information can be obtained by sending through

information request forms to the various organisations. The following section outlines the principle components of these information requests and suggests possible information that should be requested.

Internal information and pre-survey questionnaires

Pre-survey questionnaires (PSQs) are questionnaires that are sent out prior to the site investigation phase of the audit. It is an important document that enables a considerable amount of basic information to be collected prior to the investigation. It helps to reduce the amount of time spent on a site by an auditor, time which is both costly to the auditors and the auditees. A PSQ should be seen as supplementing the audit protocol (a subject discussed below) used during the site visit. The PSQ has a number of uses:

(1) It enables basic information about the site (*e.g.* plans of the site layout, aerial photographs, copies of permits and licences) to be prepared and either sent to the audit team in advance, or held on-site in expectation of the on-site review or inspection.
(2) It gives the auditor an understanding of the environmental concerns and issues at the facility, and provides a basis upon which to modify the intended standardised audit protocol.
(3) It prepares facility staff for the audit, giving them an understanding of what will be expected of them in terms of the type and level of detail of information required. It also helps them to adjust to the concept of the audit.
(4) It provides an initial focus of discussion in the interviews between the audit team and the facility staff.
(5) It enables the audit team to identify the key facility personnel who need to be interviewed during the course of the audit.

The PSQ should address a number of areas:

(1) *General overview*. The PSQ should look at the site's history, the local environment and community in which the facility is situated and the site's layout and document any previous environmental audits or reports.
(2) *Facility operation*. The PSQ needs to examine the operations, activities and processes undertaken at the site, and their scale. The

products produced at the site should be identified.

(3) *Management systems.* The PSQ should examine the site's management systems and solicit details of all environmental responsibilities, procedures and emergency plans and strategies.

(4) *Legal issues.* The PSQ should outline the regulations and legislation that affect the facility, and should identify any permits, discharge consents or licences held by the facility.

(5) *Waste outputs.* The PSQ should determine any atmospheric emissions, aqueous discharges to water courses or sewers and any solid wastes produced by the facility.

(6) *Storage.* The PSQ should detail how raw materials are stored on site and transported around the site and, if appropriate, information should be provided on transportation systems for raw materials and products onto and out of the site. This may include reviews of pipeline systems, railheads, conveyors, canal-side loading areas, marine jetties and road haulage systems.

(7) *Environmental concerns.* The PSQ should attempt to perceive any of the environmental concerns of the facility staff, and should look at the relationship between the facility and the local community (*e.g.* by obtaining information about the numbers of any public complaints).

Each PSQ sent out should be accompanied by an introductory page. This page should outline the purpose of the questionnaire, the likely benefits of the audit to the organisation and to the respondent, the confidentiality of any of the respondents replies, and the fact that the PSQ and the entire audit has the backing of senior management.

Despite the benefits gained from using a PSQ, there are also a number of potential problems that need to be addressed. One of the main problems with using a PSQ is that the returned questionnaires are often badly completed, containing incorrect or insufficient information. This occurs where the auditees do not fully comprehend the questionnaire or where the facility does not conform to the standardised structure of the questionnaire. In certain circumstances, the PSQ may not even be returned.

In order to help prevent these problems there is a need to ensure that all of the questions used in the PSQ are explicitly expressed, leaving as little room as possible for misinterpretation. A contact address or number should be provided in case the respondent encounters any difficulty with the PSQ. Reminder letters should also be sent out after a set date if replies are not forthcoming. The audit team should run through the completed forms with the facility management as soon as the on-site phase of the audit begins, in order to resolve any potential misinterpretations.

A sample copy of a PSQ is contained within Appendix 5.

External information and information request forms

There is a wide range of regulatory and non-regulatory bodies in the United Kingdom which hold various amounts of information which may be essential when compiling an audit. Similarly, because of recent developments in the provision of environmental data, there are a number of commercial organisations which, either from their own data sources or as value added resellers of regulatory data, are able to provide important site-specific information.

During the pre-audit stage it is important that the auditor contacts the relevant regulatory authorities and/or commissions data searches as early as possible, as it can in some cases take up to two months for a competent authority to provide the information requested.

It is therefore advised that an information request form (IRF) be drawn up for each specific organisation that is to be approached. This form is essentially a list of information that the auditor requires about the site. It is useful to contact the regulatory body by telephone prior to sending the IRF, so that the actual individual responsible for the site can be contacted directly. Failure to do this may result in delays in obtaining information as the IRF may be passed through several departments prior to it landing on the right desk, if indeed it manages to arrive in the correct place at all. In cases where time scales are short (*e.g.* as is the usual case in pre-acquisition audits) it is useful to fax through the IRF. Although contact should be made as early as possible, there is no problem in contacting an organisation at a later stage if information (such as the presence of petroleum tanks) is discovered during the course of the site visit.

When making a request for information to either a regulatory body or a data provider in relation to a given site, it is essential that a location map and a grid reference are also sent through at the same time, as this ensures that the authority/data supplier knows precisely where the site is and the precise geographic boundary of the site being audited. Land registry plans used by solicitors and surveyors are useful for this purpose as they show the site with a discrete legal boundary and are reduced to fit an A4 sheet of paper.

The actual organisations that need to be contacted (and in the case of data providers the required information) depends very much upon the nature of the site being audited, the type of audit being undertaken and the processes and substances that are used on site.

In a pre-acquisition audit the bodies that tend to be contacted are the pollution control, groundwater, abstractions licensing and waste regulation departments of the Environment Agency, the local environmental health department, the local planning department and the local sewerage undertaker. The type of information available from these bodies is described below. In the case of the Environment Agency, note that there are a number of departments which may need to be approached. The various departments can no longer be approached individually but rather via the Agency's Customer Services Department, or data can be obtained from one of the Agency's value added resellers. The Customer Services Department deals with all information requests for a site centrally and can collate all the requested information together. Although feeding all the information through to Customer Services can make life a little easier for the auditor (as it means that there is only one person to deal with), inevitable delays are experienced, which can often be frustrating and sometimes costly if tight commercial deadlines are exceeded.

In addition to the organisations listed above there are a number of other authorities (and related data sets) which can also be contacted if thought to be relevant to the site and the operations undertaken there. The HSE should be contacted if it is thought, for example, that the site falls within the ambit of the Control of Industrial Major Accident Hazardous Substances Regulations 1984 (SI No 1902, as amended) or the Notification of Installations Handling Hazardous Substances Regulations 1982 (SI No 1357), and the petroleum licensing authority should be contacted if it is thought that there are underground petroleum storage tanks present beneath the site. The petroleum licensing authority (usually within the local fire brigade or Trading Standards Department) can provide information on the size, age and location of any petroleum storage tanks, whether any such tanks have been decommissioned in the past, and whether there have been any significant spills on the site or leaks from the tanks.

The following outlines suggest information which should be established for a site (either by direct consultation with the given authority, or by requesting this data from a data supplier).

Information from the Environment Agency

Pollution Control/Water Quality Department
- Details of any known water pollution incidents related to the site, and details of any resulting prosecutions and fines
- The location and quality of any nearby controlled surface waters
- Details of any discharges from the site into controlled waters
- Whether the site has been visited by the agency.

Groundwater Protection Department
- Details of any aquifers located beneath the site such as its composition, its vulnerability etc
- Details of any source protection zones within which the site may lie.

Abstractions Licensing Department
The Abstractions Licensing Department holds information on all surface and groundwater abstractions. When undertaking an audit a radial search can be commissioned of all abstraction licences within a given distance from the site (a distance of 2 km is usually used). For each licence within the target area the following information can usually be obtained:

- Licence number
- Licensee details
- Use of abstraction (*e.g.* public, agricultural or industrial supply)
- Quantity of abstraction
- Source of abstraction.

Waste Regulation Department
- Details of whether the site has ever been infilled, landfilled or used for waste treatment purposes
- The location, and nature, of any landfill sites within 500 metres of the site
- Whether the site has ever been the recipient of a Waste Management Licence or a Waste Carriers Certificate (obtain copies if they exist)
- Whether there have ever been any problems with the site in regard to waste issues (*e.g.* whether it has been fined or prosecuted for environmental infringements).

Local authority – environmental health department

The Environmental Health Officer with responsibility for the area within which the site is set should be able to advise, among other things, whether:

- the site is subject to any Part B processes prescribed for air pollution control under EPA 1990 (a copy of the consent should be obtained if possible)
- the site was included in the Derelict and Despoiled Land Survey of England and Wales
- the site has been used for mineral workings
- the site has been subject to any reclamation or remediation
- complaints have been made against the site in respect environmental nuisances (*e.g.* noise, odour etc)
- there have been any contamination problems with or near the site
- there are any pre-licensed landfill sites within the site's vicinity
- the authority has just cause to believe the site to be contaminated.

Local water company – trade effluent department

The trade effluent department of the local water company should be contacted to determine whether the site has any trade effluent discharge consents. If such a consent exists it should be determined what the terms of the consent are and if there have been any known breaches of the consent.

Chapter 5

Audit Methodologies and Working Papers

5.1 **Introduction**

It was mentioned at the start of this book that auditing should be a systematic, documented, and periodic evaluation. This chapter looks at two of these aspects: the need for the audit to be systematic and documented.

A systematic audit is one that is comprehensive and covers all aspects of a site and the operations undertaken thereon. In order to be systematic a common method should be pursued. In environmental auditing, methods which have been developed predominantly revolve around the use of audit protocols. This section will therefore examine the range of protocols that can be used in auditing. These tend to fall into two categories: checklist protocols and questionnaire protocols. Protocols can also be enhanced by the use of questionnaires such as PSQs (described in Chap 4 –see p 98) and Internal Control Questionnaires; these are discussed here as well.

In regard to the documentation of an audit, a rigorous procedure needs to be established so that all information collected on a site is written down and stored in an accessible and workable format. With this in mind, working papers can be developed and these are discussed later in this chapter.

5.2 **Audit protocols**

The majority of environmental audits should be accompanied by written documentation which is used as a guide to the field work associated with the environmental audit. This documentation is usually collectively referred to as the "audit protocol", although in some situations it may be referred to as the "audit work programme", the "audit review programme", the "audit manual" or the "audit guide.

Audit protocols are usually prepared in advance of actual site visits. Protocols tend to be standardised, especially if the audit is part of a larger audit programme. Audit protocols have a number of functions, and are used:

(1) *To provide guidance to the audit procedure.* Audit protocols should list every step to be undertaken during the course of the field work stage of the audit.
(2) *As a tool to assist audit planning.* As an audit protocol lists the steps to be followed during the audit, it can be used to plan the required field activities, to assign individual team member roles and responsibilities, to allocate financial resources and to plan the overall time scale of the audit.
(3) *As a record of change in the scope of the audit.* The audit protocol can be used to show changes made to any protocol during the course of the whole audit process. As protocols are adapted throughout the audit process, the final protocol can be compared to the initial, standardised protocol upon which it was based. This will identify all changes in the scope of the audit.
(4) *As a structure for the working papers.* The protocol forms the basis for organising the audit team's working papers.
(5) *As a basis for the critique of the audit.* Upon the completion of the on-site audit stage, the protocol provides a means of reviewing the audit, and the means of ensuring that all relevant steps have been accomplished. The protocol helps therefore to ensure consistency and quality in the audits.

Audit protocols can vary widely in their format and specificity depending upon the audit's scope and the objectives, the audit team's experience and the nature of the organisation or facility being audited. The majority of protocols are either checklists or questionnaires.

Checklists

There are a number of different checklists that can be employed as audit protocols. These checklists all vary in their detail and sophistication, and in their reliance on the skills of the individual auditor.

Basic checklist

A basic checklist essentially consists of a list of the key steps in the audit process detailing them in terms of the action required on behalf of the auditor (see Fig 5.1). The protocol should allow room for identifying assignments, suitable brief comments and a working paper reference that relates to the relevant documentation produced for the specific procedural step. The advantages of using such a simplistic protocol are its:

- Flexibility as a planning tool
- Value in focusing audit resources
- Value in organising working papers
- Explicit nature that details what the auditor should do and how he should do it, which provides consistency within the audit programme
- Use in reviewing the audit's progress, and in helping to ensure that the quality of the audit is satisfactory.

Topical checklist

A topical checklist consists of a list of all the topics to be covered during the course of an audit. As a straightforward list this type of protocol relies heavily on the auditor's experience and knowledge, as it presumes a great deal by not giving any details about how to undertake the auditing of each of the topics. This makes it an inappropriate protocol for inexperienced auditors. On the other hand, the protocol is short and easy to use for more experienced auditors, who have the knowledge and experience to be able to audit each of the specific areas. To such an auditor, the protocol simply represents an aide-mémoire and ensures that all areas have been considered.

An example of part of a topical checklist is given in Figure 5.2. This small section represents the solid and hazardous waste section of a topical checklist.

Detailed checklist

A detailed checklist is an in-depth attempt to list everything that should be known and undertaken by the auditor. Thus it provides the auditor with an outline of all regulatory requirements and basic responsibilities within the facility. It essentially holds the hand of an auditor, providing

Audit Steps	Auditor(s) assigned/ general comments	Working paper reference

1.   Examine all relevant documents prior to visit to determine scope:
    • Facility layout
    • Regulations; Federal, state, local
    • Applicable permits
    • Policies and procedures; corporate, facility
    • Operating manuals
    • Paper and report flow
    • On-site systems for handling, storage and
    disposal of solid and hazardous wastes		

2.   Document the scope of the audit in the working audit papers:
    • Time period under review
    • Compliance auditing versus hazard auditing		

3. Develop an understanding of the facility's internal management controls through completion of the internal controls questionnaire or through other means such as discussions with facility manager, environment manager, etc.		
4. Document your understanding of the facility's internal control in flow chart or narrative form, showing responsibilities for action and record-keeping for solid and hazardous wastes.		
5. Tour the sites to be audited to reconfirm and better understand the facility's internal controls.		

6.   Testing:
    • Identify different types of tests and describe in working papers. Prepare a schedule of key tests.
    • Test selected transactions to confirm paper flow. Document in working papers.
    • Develop a confirmation testing plan designed to verify compliance.
    • Test selected transactions for compliance.
    • Obtain instructions for labelling and dating wastes received into storage. Inspect labels and dates on a selected sample of wastes in storage. Document agreement between inventory records and wastes in storage
    • Record results in working papers.		

Figure 5.1: Example of a page from a basic checklist protocol (based on ICC, 1991).

Solid and Hazardous Waste

- Waste characterisation:
- Manifest system:
- Training programmes:
- Inspections:
- Preparedness and prevention:

Figure 5.2: Section of a topical checklist protocol.

an excellent medium for training inexperienced auditors. With well designed, detailed protocols an inexperienced auditor should be able to conduct an audit with relatively little supervision from the audit team leader. This reduces the pressure on the leader as there is some guarantee that the auditor is receiving a certain level of guidance. Detailed checklists also tend to be used in situations where the audit team is unaware of the regulations that govern the operation of the process.

Detailed checklists provide a good means of assuring audit quality as they can be used to review an auditor's work, checking that it has all been undertaken properly. In some cases, however, these checklists may become too long and unwieldy to be used effectively. Also, there may be a problem with collecting unnecessary information. Detailed checklists require specific areas to be examined during the course of the audit, but in some instances sections of the protocol may not be relevant to an individual facility, and useless information may be collected. This may constitute an inefficient use of resources.

5.3 **Questionnaire protocols**

Yes/no questionnaire

The yes/no questionnaire is one of the most widely used forms of questionnaire protocol used in auditing. Such questionnaires are usually quite long and detailed, a necessity when considering that all answers are limited to a "yes", "no", "don't know" response. The coverage of the questions has to be broad, and tends to centre on the relevant regulatory requirements covering a facility. If the questionnaire is produced in-house, then it may well cover the requirements of the company as well. As the questionnaire provides guidance on regulations (and company policy) the auditor need not have extensive knowledge of them. This is the main advantage of using such a protocol as totally inexperienced personnel can be used to apply them during an audit.

One of the major problems with the protocol is its dependence on enquiry as a means of gathering information. Little scope is left for observation and verification testing. This reliance upon enquiry does not actually determine whether compliance is being obtained. A further problem with this type of protocol is that it can become too lengthy and bulky to be used effectively.

Figure 5.3 represents an example of a yes/no questionnaire.

Question 3:
Have there ever been any leakages from any of the tanks to your knowledge?
YES / NO

Question 4:
If answer YES ... which tanks or pipework failed and what amount of fuel was lost?

Type of fuel lost?

Question 5:
Were any remedial measures necessary by:

    Owners: YES / NO

    Fire Brigade: YES / NO

Figure 5.3: Example of a section of a yes/no questionaire (EAL, 1999).

Open-ended questionnaire

An open-ended questionnaire is one that enables auditors to obtain a more in-depth response to questions, rather than a simple "yes/no" response. Thus a factual and explanatory answer can be obtained to a specific question. The approach is still one that relies heavily on enquiry, resulting in the compliance status of a facility remaining unverified by the use of this protocol alone.

Figure 5.4 represents an example of this form of questionnaire.

SECTION FOUR: SITE HISTORY

4.1 How long has the site been used for its present use?

4.2 What year did the current company occupy the site?

4.3 What was the site used for prior to its current use?

4.4 Have any contaminated land surveys been undertaken at site?

SECTION SIX: SITE DRAINAGE AND DISCHARGES

6.3 Are drainage plans available for the site?

   Append a copy of these.

6.4 Are these drainage plans up to date, and have there been any

   modifications since?

6.5 What discharges are made to the storm water drains?

6.5 Where do these discharge to?

Figure 5.4: Example of a section of an open-ended questionaire (EAL, 1999).

Scored questionnaire

Scored questionnaires are similar to the questionnaires outlined above. However, these questionnaires also attempt to measure the facility's environmental performance by scoring each of the responses. A performance score is obtained by comparing each answer against a master template. Scored questionnaires are useful as they provide management with a good summary of a facility's environmental performance, and thus allow for inter-facility comparison. If applied universally within an audit programme, they can be used to identify overall company trends and problems, allowing overall company initiatives to be developed (*e.g.* the introduction of new internal company policies).

Internal control questionnaire

Internal control questionnaires can be seen as another supplement to the audit protocol. They are a quick and effective means of obtaining information about the status and extent of the environmental controls within the facility being audited, providing the audit team with a detailed knowledge of the key controls within the facility and the individuals responsible for them. Information obtained by the questionnaire should be far more detailed than that obtained during the opening meeting. The questionnaire is usually administered to the individuals responsible for the environmental controls by the audit team leader. It is advised that the rest of the audit team are present during such a meeting, as this allows them to ask specific questions regarding their own area of interest. The questionnaire should act as the guide to the auditors through such interviews.

5.4 **Working papers**

Working papers constitute the notes made by the environmental auditors throughout the auditing process, and in particular the on-site phase of the audit. They have a number of important functions, and are used to:

(a) provide support for the audit report in the form of detailed information that backs up any conclusions or recommendations made by the audit team (this detailed information may be in the form of notes taken during interviews with facility staff, from the results of any compliance testing undertaken or from the review of any facility documentation; it is especially important that the working papers record the data or information that corroborates any findings of compliance or non-compliance);

(b) assist the auditors in the organisation of their work by providing a structured method for ensuring that all the steps prescribed in the audit protocols are completed and followed;

(c) document the rationale for any tests undertaken by the audit team, and to record any of the test results obtained;

(d) supplement the audit protocol by recording any deviation from the protocol and the reason behind such deviations;

(e) give documented evidence that the audit was conducted in a manner that conformed with the aims and objectives of both the individual audit and the audit programme;

(f) provide the audit team leader with the basis for assessing the quality of the work of each of the individual team members; and

(g) provide background information and a future reference source for other forthcoming audits.

Content

Working papers are used in both the pre-audit and on-site activity stages of the audit process. Their content reflects the different nature of information gathering at each of these stages. In general, working papers collect three main types of information.

Initially they will record general background information about both the audit and the facility to be audited. Such information regarding the audit, namely its scope and the protocols to be used, is usually collected in the pre-audit stage. Background information about the site, documented in the working papers, is usually obtained from the results of the PSQ (if used) and through an initial opening meeting. The working papers should document the completed PSQs and record the transactions of the opening meeting. This involves listing all the personnel present at the meeting and recording all of the information obtained during it. Observations made during the inspection of the facility (*e.g.* the location of chemical storage vessels) should also be recorded in this section of the working papers.

The working papers should also record the audit team's understanding of the management of facility operations. For each operation, the working papers should as a minimum record the individuals with responsibility for the area, the way in which the operation is managed, what documentation is kept for it and how often the activity is operational (*i.e.* how often does the process run? what are the shift patterns? etc). The methods for recording such information in the working papers consists of structuring a flow chart of the process and its associated management system and/or by writing a detailed description of the activity.

Finally, the working papers need to record and document the process by which each auditor gathers audit evidence and information for the completion of each audit protocol step. At each stage of the audit protocol the working papers should record:

(1) All information gathered in the protocol step and its related source.
(2) Any tests conducted during the course of the respective protocol step. For each test conducted, it is important that the working papers record:
    (a)   the objective of the test;
    (b)   the population, subject or topic being tested;
    (c)   the rationale behind why the test should be taken and why that specific test;

(d)   any sources of potential bias in the test;

(e)   the size of the test sample and the justification for that size; and

(f)   the results of the test.

(3) Any key observations made during the protocol step.

(4) Any departures from the audit protocol and a justification of the departure.

Review

Working papers need to be reviewed by both individual team members and by the audit team leader. Each individual auditor should review his working papers on a daily basis for a number of reasons. First, the auditor needs to identify any initial conclusions and findings for each area covered and identify whether any further information is required. Secondly, any conclusions that are reached by the auditor should be

---

Format
- Each working paper page is clearly labelled with the protocol step.
- The sources of information are clearly identified.
- All exhibits are referenced in the working papers.
- Each page is sequentially numbered, initialled and dated.
- Cross-outs are initialled; postscripts or afterthoughts are written in a manner that provides appropriate context.

Content
- Each protocol step was completed in accordance with the instructions provided.
- Any departures from the protocol are described and explained.
- A description of actions taken to complete each protocol step has been documented.
- An understanding of how the facility in managing the items under review has been documented.
- The conclusions reached as a result of testing have been documented.
- All audit findings have been clearly identified.
- All findings identified in the working papers have been included in the audit exit meeting discussion sheet.

Figure 5.5: Checklist of review working papers

documented and included in the working papers as interim summaries. These interim summaries provide the auditor with a vehicle to outline the findings of each of the protocol steps. Thirdly, the auditor needs to check that each protocol step followed has been properly completed and that there are no outstanding areas requiring attention. This review should also check that the working papers are all in the correct format and order. Figure 5.5 provides a checklist that can be used to assist the auditor in such a review.

The audit team leader should also review the working papers of each individual auditor to ensure that the appropriate standard is being reached, and the working papers are of a sufficient quality. Such a review is an important means of ensuring the quality of an audit.

Characteristics

There are a number of rules that should be followed when preparing working papers. These rules are widely reproduced in audit literature:

- Working papers should be legible
- Working papers should be initialled and dated
- All sources of information should be written down
- All parts of the working papers should refer to a relevant stage in the audit protocol.

Each section should:

(a) be summarised by an interim summary which should draw together the auditors conclusions on that section;
(b) provide exhibits (usually photocopies of the relevant documentation) to support all positive and negative conclusions;
(c) fully explain all systems that are tested and should describe any sampling methods used;
(d) not have any sections that are incomplete or which leave areas of importance unanswered; and
(e) be read over frequently by the auditor to ensure that they are sensibly written.

Retention

Working papers should be retained for an appropriate period of time as they provide the basis supporting the audit findings. Thus in situations where a particular conclusion of the auditors is being contended, there will be a need to refer back to the working papers to ensure that the conclusion is based on fact and substantiated by evidence. The length of retention varies but should be approximately four years, or until the audit is repeated.

Action lists

Information not classified in the working papers or pending supply should be detailed and a list should be constructed to accompany the working papers. This list should set out "additional requests for information". This list can be discussed with site management and agreed, and the information can then be sent on to the auditor after the team leaves the site if necessary.

Chapter 6

Identifying Environmental Effects

6.1 Operations

The assessment of an operation's environmental effects can be undertaken at the time of the site visit. Upon completion of the site visit, information derived from this function can then be correlated with other information derived from additional information sources (*e.g.* historical maps, planning records etc – see Chap 4). One of the key methods of assessing the environmental effects of an operation is first to establish the principal operational factors that are related to the site. These may ascertained from, but are not limited to:

(a) *the health and safety record for the site:* information related to the occurrence of accidents, the presence of an accident report book and notification from regulatory authorities;
(b) *emergencies:* information related to the presence of a management system or contingency plans;
(c) *the general physical condition of pollution abatement equipment:* an assessment of this equipment and whether it meets the regulatory management programme and/or any maintenance programme established for such equipment should be established;
(d) *training and management:* investigations into whether training exists for the site's operational management and staff for health, safety and environmental issues should be undertaken.

When attempting to establish the environmental effects associated with an operation undertaken on a site, it is important to draw a distinction between an actual environmental problem (a problem area) and an area in which there is a potential environmental problem (a danger area). Possible indicators of problem areas may include:

(a) signs of contamination of surface materials, including water (*e.g.* water discoloration, the presence of an oily film etc);
(b) signs of contamination of land or soil (*e.g.* stressed vegetation, discoloration of foliage);
(c) evidence of leakage (*e.g.* staining associated with an above-ground storage tank);
(d) unmonitored, unclear waste streams from production to disposal of waste, including transport;
(e) evidence of buried waste (*e.g.* waste materials in abandoned pits, inadequately stored chemical drums);
(f) evidence of uncontrolled discharges (*e.g.* open-ended pipes, unconnected drains); or
(g) evidence of fires, explosions or other physical damage pertaining to a process or storage area.

Alternatively, possible indicators of danger areas may include:

(a) the existence of contaminated land, or land which has been cleaned up (*e.g.* from previous poor waste management and disposal, leaking pipes);
(b) waste treatment, generation, disposal and storage;
(c) the existence and condition of storage tanks and drums;
(d) the existence of hazardous substances and materials;
(e) the existence of drums and tanks containing liquid materials, unbunded above-ground storage tanks, inadequate secondary containment;
(f) the existence, storage and use of radioactive materials;
(g) the detection of gases or fugitive emissions (*e.g.* methane, hydrogen sulphide, hydrogen cyanide, radon);
(h) the existence of stacks and vents indicating atmospheric emissions;
(i) the existence of discharge points (*e.g.* unconnected drains, above-ground storage tanks, haphazard piles of materials);
(j) poor housekeeping; or
(k) indications of a lack of regulatory compliance.

During the identification of potential environmental effects associated with an operation it is important to be vigilant. Manhole covers, drains, storage areas and pollution abatement equipment should be checked and the auditor should be aware of any noxious odours; open, unprotected or unlined pits or lagoons; the absence of bunding for above-ground storage tanks; and/or inadequate housekeeping. When assessing the condition of pollution abatement equipment it is important not only to consider its present condition, but also the likelihood of the equipment requiring upgrading to meet future requirements under IPC, LAAPC or IPPC.

6.2 **Operational impact areas**

When undertaking an audit, depending on the scope that is required, the auditor may be required to assess the potential environmental effects of a number of operational factors associated with the target organisation. Areas that may be encompassed within this assessment include:

- Product design
- Packaging
- On-site storage
- Process design/operation
- Transport/distribution
- Site management
- Energy use/efficiency
- Water use/efficiency
- Sources of raw materials
- Emissions and discharges to all environmental media including discharges of thermal energy, noise, odour, dust, vibration and visual impact (including light intrusion)
- Creation of major electro-magnetic fields
- Waste management.

Product design

If the company/organisation manufactures a product or if the company has specified a product as part of a process, the auditor may be required to consider the environmental effects of the product. The examination should not just consider the product's impact solely in terms of final disposal, but should also include the raw materials or components and the actual processes involved in making it. The examination should not solely consider the legal obligations attributable to the industry, but should also look at the wider issues such as pressure from environmental groups, stakeholders and customers. For example; the use of tropical hardwoods within an industrial process, although not covered by specific environmental legislation, may not be considered to be best practice within the industry. The examination may include the following steps:

(1) Assessment of how product design influences the process and how the process determines the design.
(2) What are the determining factors in the design of the process (*i.e.* customer demands, specification requirements etc)?
(3) How does environmental legislation affect the product design, and are legal requirements likely to be more stringent in the future?
(4) What are the energy and materials used in the process, and is there scope to reduce the quantities used?
(5) What are the final disposal options for the product once it has come to the end of its "working life"?
(6) Is the company required to take back the end product or packaging for ultimate disposal by the customer?
(7) Is there scope for building in durability/recyclability within the product? Could a competitive/marketing advantage be gained from such an undertaking?

Packaging

The Producer Responsibility Obligations (Packaging Waste) Regulations 1997 and the EU Packaging Directive have set targets for recovering and recycling packaging materials. Following prompting by the DETR, the major UK retailers have agreed to recover 58% of packaging (of which 66% must be recycled) by 2000. Correspondingly, an auditor may be required to look at the effects associated with the packaging materials used by the company/organisation being audited. The auditor should look at the following information:

(1) How much packaging does the company use in a year?
(2) How much is recycled material and how much is virgin?
(3) Is there scope to reduce, reuse or recycle?
(4) Do the provisions of the 1997 Regulations apply to the organisation?
(5) Ascertain whether goods always have to be packaged.

On-site storage

When undertaking an audit, an auditor should be aware of the environmental effects associated with the storage of raw materials on a site. When identifying the risks associated with such storage, an auditor should identify:

(a) the individuals responsible for the storage of materials and ensure that they are well informed about the legal requirements of such storage, especially in relation to the Control of Substances Hazardous to Health Regulations 1988 (SI No 2533 and the Control of Industrial Major Accident Hazardous Substances Regulations 1984 (SI No 1902, as amended);
(b) whether the storage site is suitable in terms of:
   (i)   slope and permeability of land;
   (ii) location of drainage systems;
   (iii) location of centres of population; and
   (iv) location of sensitive resources (*i.e.* rivers etc);
(c) whether there are any preventative measures in place to minimise the risks associated with spillages;
(d) whether other preventative measures can be taken in the storage area;
(e) whether the procedures in place are appropriate for the individual materials and whether they meet the appropriate regulatory requirements; and
(f) whether there are any fugitive emissions associated with the storage of materials.

The risks associated with on-site storage are especially important given the recent provisions included within the Groundwater Regulations 1998. On-site storage is important in relation to the degree of risk associated with a site. For example, in 1994 a fine of £7,000 was imposed for causing a discharge of polluting matter to the Manchester Ship Canal in breach of the Water Resources Act 1991. A burst tyre on a forklift truck had resulted in the puncture of a chemical container (ENDS Report 236, 1994).

Process/design operation

Many businesses develop in an ad hoc manner as they grow and respond to normal commercial pressures. As they expand, new machines are incorporated within the existing system which can lead to inefficiency. For example, an aluminium recycling company replaced its old remelting equipment with a new furnace designed to be both energy efficient and environmentally clean. The overall cost savings including improved yield and the purchase of lower quality feedstock was estimated at £348,000 per year, with a payback period of 11 months. Correspondingly, within the scope of a specific audit an auditor may be required to become familiar with the processes undertaken on a site, talk to individuals on the site such as operators, technicians and process engineers, and then consider whether:

(a) there is scope to minimise the consumption of energy, raw materials or water;
(b) there is any monitoring of efficiency undertaken for a process;
(c) there are areas within a process in which a new piece of equipment could potentially improve efficiency and have a beneficial effect on the environment;
(d) there are any fugitive emissions associated with the process that could be reduced;
(e) process chemical substitutions would replace a potentially harmful chemical with a more benign chemical (*i.e.* water-based instead of solvent-based inks in the printing industry).

Transport and distribution

Transportation is an intrinsic part of most businesses operations, with even service sector companies utilising a mobile sales force. The bigger the organisation, the greater the potential effect. Correspondingly, an auditor should consider the following:

(1) Does the company have a car policy (*i.e.* in relation to mileage allowance, does the allowance encourage travel)?
(2) Does the company monitor mileage and fuel consumption?
(3) Does the company encourage car sharing?
(4) Do drivers receive training?
(5) Are commercial vehicles regularly maintained? Poorly maintained vehicles are less efficient and more polluting (*e.g.* poorly inflated tyres increase fuel consumption by 5%).
(6) Does the company monitor the distribution schedule and is there room for improvement?

(7) In relation to warehousing activities, is there scope to reduce heat loss and noise pollution?

Site management

Site management should be concerned with reducing the site risks, improving the visual amenity, the current land usage, site location and the services that are bought in. When an auditor is required to assess the site management, the following aspects should be considered:

(1) Is there effective land usage (*i.e.* what percentage of the site is derelict and not in use, what is the general state of repair)?
(2) Identify those areas of the site which are environmentally important – are they appropriately located within the site's perimeter?
(3) Are the staff that have responsibility for the site's management sufficiently knowledgeable in relation to the risks associated with the site?
(4) Does the site's basic infrastructure contribute to the site's real potential environmental effects?
(5) Does the site have in place any emergency procedures (*i.e.* a contingency plan)? Note that this is mandatory for Control of Industrial Major Accident Hazardous Substances Regulations 1984 (SI No 1902) sites.

Energy use, water use and efficiency

Research has shown that the majority of businesses can reduce their energy bills by 10% without any capital expenditure. Old equipment, poor maintenance and lack of basic insulation all contribute significantly to the operation's energy efficiency. Correspondingly, estimates relating to the investment in control devices that limit the usage of water on a site (*e.g.* flow regulators, tap flow reducers and smaller diameter hoses) can save 25% of water usage, with a pay back period estimated at 12 months or less. Correspondingly, when assessing the efficiency of an operation, an auditor should consider whether:

(a) there is scope within an operation in which water can be reused, rather than flushed away;
(b) the company monitors any discharges in relation to environmental media;

(c) poor housekeeping or operational factors contribute to the level of effluent being discharged from a site and whether there is scope for improvement; and

(d) there are areas of the site which are poorly maintained, such as leaking pipes, which may be contributing to water loss.

Sources of raw materials

It should be an aim of any business to reduce the use of natural resources, although it should be understood that the use of raw materials for processes is essential. Correspondingly, a degree of common sense needs to be applied to any assessment of the use of raw materials and, where practicable, off-the-shelf components should be sourced from manufacturers with the least possible environmental impact. The auditor should consider whether "supplier" auditing is necessary and appropriate, depending on the size of the facility being audited.

Emissions and discharges

From the assessment of the operations undertaken by an organisation, or from the assessment of an individual process, the auditor will be made aware of emissions and/or discharges which are related to that process/operation. In addition to emissions directly related to the process, such as combustion, fugitive emissions should also be considered, such as the loss of solvents from open paint cans. The auditor needs to establish whether:

(a) the company/organisation monitors the effluent quantity and volume

(b) there are any regulatory requirements associated with emissions and whether the site is in compliance;

(c) there is scope for the emissions to be reduced;

(d) the site has problems with respect to noise or odours, who deals with them and what steps are being taken to reduce the problems; and

(e) there is a mechanism in place to deal with complaints from neighbours and whether records are kept of complaints.

Waste management

The control of waste has received some widespread attention from both legislation and the media. One of the principal focuses in assessing the waste management activities associated with an operation or organisation is to establish how much waste is created and who has the responsibility for its management and ultimate disposal. Correspondingly, an auditor needs to consider the following:

(1) What wastes are produced by the various operations undertaken by an organisation?
(2) Is the organisation in compliance with its duty of care?
(3) Is there scope for the reduction in the waste produced by the various operations (*i.e.* is it a matter of better housekeeping or changes in the process plant)?
(4) Establish the waste trail and who deals with the waste at various stages from creation, collection, storage, treatment to disposal.
(5) Are there clear, basic procedures in place to deal with the waste management on the site, including labelling, storage, handling, emergency procedures and proper disposal?
(6) Is there scope to store similar wastes together rather than mixing different types of waste, since segregation of waste is often less expensive and facilitates recycling.

6.3 **Process diagrams**

Often processes and operations undertaken on a site are both complex and confusing. It is often not possible to visually identify waste streams or other possible environmental problems associated with a process. In these instances it may be useful to ask for a process diagram from the site management. The process diagram should detail the specific processes carried out for each operation undertaken on site. From this documentation, the auditor should be able to break down the complete operation into smaller subdivisions, which in turn can be assessed for potential environmental effects.

When interpreting a process diagram the auditor should look at the flow of materials through an organisation or unit, with the inputs and outputs identifiable at each stage. The inputs could include:

- Energy
- Raw materials
- Water, and/or
- Other resources (*e.g.* cooling mediums such as CFCs).

Correspondingly, outputs could include:

- Products and by-products
- Solid waste
- Liquid effluent
- Emissions to atmosphere
- Associated noise, odour, dust and vibration, and
- Visual impact.

When assessing the environmental effects of a process, the emphasis may be on the resources used and the output emissions of the process.

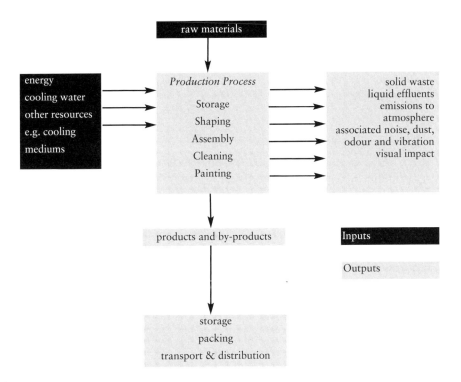

Figure 6.1: Example of process diagram.

Data obtained would therefore need to consider:
- Amount of energy used
- Types of solid and liquid waste produced
- Quantities of raw materials used, and
- Discharges to water and emissions to air.

When reviewing processes for their effects, account must be taken of both the normal and abnormal operating conditions.

Monitoring suppliers' environmental performance

When undertaking an audit, it is often necessary to undertake an investigation into the environmental performance of the company's suppliers (for both services and raw materials), since this may have a direct bearing on the subject company's environmental performance, or any environmental effects associated with the site's operation which utilise external suppliers. To this effect, inquiries should be made relating to the following:

(1) Does the supplier have an environmental policy or environmental management system?
(2) Is the supplier aware of the regulatory requirements relating to its operations, and if so does it achieve these requirements?
(3) Does the supplier have the required environmental consents, permits and authorisations?
(4) For any permits, consents or authorisations which have been granted to the supplier, whether conditional or not, is the supplier in compliance?
(5) Has the supplier been the subject of a civil or criminal or administrative action in respect of any environmental infringements?

Chapter 7

Reporting and the Audit Process

7.1 Introduction

Reporting is an action that is likely to occur at a number of stages during the auditing process. Consequently there are a number of different reporting mechanisms that can be used. These include informal reporting on a day-to-day basis with facility management, the formal reporting of results in the exit/close out meeting and the formal reporting of results through the main report. The audit team and, specifically, the team leader are responsible for reporting to a wide range of people and the tone and content of the report will therefore vary.

7.2 Oral reports

The main phase of oral reporting occurs between the auditors and the auditees throughout the on-site phase of the audit. This oral reporting is informally based and optimally should happen at the end of every working day. The purpose of this frequent oral reporting is to:

- Facilitate a joint understanding of the relevant points between the auditor and auditee
- Create a sense of participation on the auditees' part, encouraging an open and co-operative attitude towards the audit
- Develop the awareness of the facility's staff of the exact nature of any deficiencies, providing a basis for prompt response
- Enable any problems or misunderstandings to be discussed and subsequently resolved.

At the end of the on-site phase, an exit or close-out meeting should be held. This exit meeting involves the formal oral communication of the audit findings to facility personnel. It is an extremely useful exercise to conduct, as it is the first time that the auditors' conclusions will be properly discussed outside the audit team, and it enables an immediate response to be obtained. Auditors need to prepare such reports with care and should focus on the needs, interests and knowledge level of the respective recipients (ICC, 1991).

7.3 **Formal written reports**

A formal written report draws together the final conclusions of the audit process. This report is an important document as it is the primary means of communicating the audit findings to personnel not directly involved with the operation in question, and it also constitutes an enduring record of the audit procedures followed. The report should provide management with information about the facility's compliance status, provide recommendations that will serve as a basis upon which to initiate action, and document how the audit was undertaken, what was addressed and what was found (ICC, 1991).

The report's precise format and structure will vary depending upon the audit objectives and the company's management style. The format should be clear, concise, and easy to use. The report should also be relevant to the recipient's needs. This means that it needs to be produced as quickly as possible so that it provides the reader with up-to-date knowledge. The audit cannot be prepared over a period of months as, by the time the report is completed, the situation most likely will have changed. Therefore, draft reports should be produced within one week of the completion of the on-site audit, and a finalised report should be published soon after. By its very definition an audit is a "snapshot" of a facility or company at a particular moment in time (*i.e.* the time at which the specific audit was carried out).

Although the structure of an audit is to an extent dependent upon the type of audit that is being undertaken, there are a number of sections that should be included within an audit report.

A contents page. This should appear at the front of the report.

Executive summary. The summary should introduce the audit and its main findings and recommendations.

Introduction. The introduction to the audit should outline the audit scope, the report structure and details of the commissioning party.

Description of the site. An (illustrated) description of the site should be incorporated into the report. This should indicate the various buildings present on the site and the various operational areas present. This section can also be used to describe the site's past uses and operational history. It usefully can be augmented by provision of a site plan showing the site in relation to its immediate environs and neighbours. On more complex industrial sites it may be further advantageous to provide detailed "plot plans" showing the equipment and processes taking place in specific areas within the facility, or to provide an overview of processes accompanied by reference to the key piping and instrumentation diagrams for the site, which should be included as appendices.

Assessment of performance against agreed criteria. This section should highlight the strengths and weaknesses of the site being audited and should identify all areas of non-compliance. The assessment should be factual rather than subjective, and must be substantiated by evidence wherever possible. In some situations it is often beneficial to assign values to the facility's performance against certain criteria. Facilities can be evaluated against a number of areas and, for example, assigned a value between 1 and 4. A value of 1 indicates poor performance, and 4 an excellent one.

Performance can be examined in a number of areas: organisation and management; gaseous emissions; aqueous effluent; energy efficiency; waste; noise; visual impact; and transport. However, this is only an indicative guide and it is recommended that companies develop their own systems. The benefits of a performance guide like the variety described above is that it is of use when comparing sites with other, similar sites. Such comparison is useful as it determines whether there are any common problems within an organisation where there is more than one site. It can also be used to evaluate a site over a period of time, showing whether there has been an increase or decrease in environmental performance.

Reference to corrective actions from previous audits. Where possible, the audit should refer to any previous audits of the site, highlighting previous compliance, recommendations made at the time and progress achieved since then.

List of action items and recommendations. The report must outline all specific areas where action is required. Recommendations should be made where appropriate concerning improvements that could be made.

Each of these recommendations should be accompanied by an estimate of the cost and requirements of the associated action, the resources needed, the optimum time for introduction and how its implementation can be monitored.

Priority ranking. It is often useful for the auditors to rank required actions in terms of their priority. Priority can be determined by the environmental risks involved and/or the environmental desirability of the action or by the cost of the modifications needed. Prioritising actions assists management in its preparation of an action plan. Three levels of priority are detailed below.

Priority one relates to situations where there is a high probability of an environmental incident occurring and urgent action is required, or where the company is exposing itself to significant liability.

Priority two relates to situations where there is a more remote chance of an incident but which still requires attention in the near future.

Priority three relates to situations where there is no imminent danger but where the company may benefit through reducing its costs or by improving its employee, customer or public relations.

In addition to assigning a priority to an action it is also useful for the action's cost implications to be outlined.

Information regarding the audit team. The report should detail the experience and qualifications of each member of the audit team, and should describe their relationship to or position in the company.

Limitations on independence. Any restraints on the independence of the audit team must be identified in the audit report.

An International Standard for undertaking environmental auditing (see Chap 10) has recently been developed. This standard, BS EN ISO 14010, details what audit-related information should be included within an audit report. These include, but are not limited to:

- The identification of the organisation audited and of the client
- The agreed objectives and scope of the audit
- The agreed criteria against which the audit was conducted
- The period covered by the audit and the date(s) the audit was conducted
- The identification of the audit team members
- The identification of the auditee's representatives participating in the audit

- A statement of the confidential nature of the contents
- The distribution list for the audit report
- A summary of the audit process including any obstacles encountered
- The audit conclusions.

7.4 **Action list**

Information not classified in the working papers (see pp 111-115 above) or pending supply should be detailed and a list should be constructed to accompany the working papers. This list should set out "additional requests for information". This list can be discussed with site management and agreed, and the information can then be sent on to the auditor after the team leaves the site if necessary

7.5 **Criteria for reporting**

Reporting during the audit process, both oral and written, must be of sufficient quality and should meet certain criteria. Reports must be accurate in all respects. Management requires a reliable report and so all findings and statements should be supported by evidence, and documented in the working papers. Reporting should be clear, since clarity ensures that the readers (or listeners) fully understand the auditor's views and that no misinterpretation of the finding occurs. Jargon and highly technical terms should be avoided and the report should be tailored to the intended recipient.

It is also important that reporting is concise. In busy working environments, long and superfluous reports are unwelcome, and care should be taken that only directly applicable material is included. Reports should be submitted as soon as possible so that they are still relevant to the current situation, but this should not be allowed to compromise their quality. An appropriate balance needs to be established between the manager's needs for up-to-date information, and the auditor's needs for time to write the report effectively. Finally, the tone of the report is also considered to be important, and it should be presented in a calm, courteous and thoughtful narrative.

Chapter 8

Contaminated Land, · Groundwater and · Pollutant Pathways

8.1 **Introduction**

One of the principle areas of concern are the risks and liabilities associated with contaminated land. On this basis, Chapters 8 and 9 deal specifically with contaminated land and groundwater, possible pollutant pathways and methods by which contaminated land and groundwater may be "remediated". There is an extensive range of excellent literature on how to investigate contaminated land, groundwater and the assessment of pollution pathways. For this reason, only a summary of the principle contaminative issues is provided and the reader is directed to the Bibliography for further guidance in this respect.

It is important to note that none of the legislation implies that the mere presence of contamination at a site is enough to require remedial works to be undertaken. Rather, clean-up of contamination is only required when a specified pollutant linkage has been established. On this basis, it is essential that during the auditing process, specifically due diligence, full attention is paid to any potentially contaminative risks associated with the site, combined with an assessment of any potential pollutant linkage. Increasingly, full disclosure of a site's status is being undertaken in an attempt to exclude any liability from the vendor. Similarly, it is important prior to acquisition to review all the existing information and, if applicable, that additional investigations be undertaken to fully establish the site's status. Remedial actions can then be included within the ownership transaction, with clean up being a condition of the property transaction and/or retentions being held back from the site's agreed value for the purchaser to undertake the required works.

This chapter discusses the principal sources of contamination, the issues associated with contamination of groundwater and remedial options which may be considered depending on the site-specific requirements. This is by no means an exhaustive analysis, but rather a guide to the options.

8.2 **Contaminative risks**

Explosive and flammable gases

Inflammability is related to the ease with which a material (gas, liquid or solid) will ignite, either spontaneously (pyrophoric), from exposure to high temperature (auto ignition), or to a spark or open flame. It is dependent on the relative concentration of gas in the air and of oxygen present.

If inflammable gases are concentrated into a confined space and high energy is released, an explosion will result. There are two principal gases which frequently present a risk under this category and are most likely to be encountered during assessment work: methane and petroleum vapour.

Methane

Methane is a colourless, odourless, tasteless gas which has a lower explosive limit in air of 5% and an upper limit of 15% by volume. Both natural gas (North Sea gas) and coal gas (town gas) are principally comprised of methane, which was formed during the anaerobic microbial degradation of oils and coals respectively. It is not surprising, therefore, that methane is also one of the principal components of landfill gas.

There are a number of situations whereby this gas may become a cause for concern.

Redevelopment of old landfill sites

Many industrial, commercial and even residential properties have been built on reclaimed landfill sites. The Environment Agency estimates that some 14,000 active landfill sites may pose a risk because of gas emissions. Half these sites lie within 250 metres of housing or industrial development and less than 30% have gas control equipment installed. The threat to a property is posed by construction of the building either directly over a "gassing" site or adjacent to such a site. Methane migrates laterally through the soil for a distance of 300–400 metres, although in exceptional cases it has been found to have migrated laterally much further.

Decomposition of other organic contaminants

Many former industrial sites (as well as many modern facilities), particularly in the metal industries, used large amounts of oils for lubricating machinery and products. These oils soaked into the soils underneath the site, or lay underneath storage tanks where they decomposed microbially, generating methane. In confined situations the gas can build up and can occasionally result in an explosive atmosphere. In such cases, dangerous mixtures can build up in basement areas or sub-surface voids.

Leakage from gas transmission systems

Town gas manufacture by the coal carbonisation process has ceased since the advent of cheap, available supplies of natural gas. However, the residues that remain on old gasworks sites are another story. Natural gas leaks from mains systems are fortunately rare (though dramatic when they happen). Natural gas is artificially given an odour to make its presence easily detected, as in its natural state it is odourless, unlike town gas which has a strong and distinctive odour cause by impurities in the gas. Consideration only needs to be taken where a large-scale acquisition of land includes transmission pipes, where small operating losses could conceivably damage high value horticultural crops, or where "blanked-off" underground pipework remains beneath the property but has not been properly gas-freed. It is a little known fact that gas losses in the UK transmission system are approximately 3–4% by volume on a normal operating basis.

Petroleum hydrocarbons and associated vapours

Petroleum is a complex mixture of highly volatile hydrocarbons including benzene, toluene and ethylene. When spilled on a site, it can be transported through the soil either as a vapour phase gas or as a solute in groundwater. Because petroleum vapour is highly flammable as well as being explosive, any leakages or accidental spillages are accorded high priorities. Often petroleum will be transported on the water table considerable distances before volatising and migrating vertically through the soil. This constitutes a particular risk to third parties off-site, with explosive flammable material building up in basements, cellars or sewers, or in sub-surface buried utilities such as cable runs, pipework and drainage systems.

There are approximately 14,000 operating (branded) petrol retailing sites in the United Kingdom, and many other industrial or commercial sites where petroleum products are stored above or below the ground. The cost of cleaning up a contaminated and potentially dangerous service station can range between £50,000 and £250,000, depending on the location and the extent of the contamination.

Deleterious materials

These substances (or other agents) directly attack the fabric of buildings, are hazardous to health, or may be responsible for the loss or interruption of services to buildings. Many previous industrial sites (steel works, chemical works, gas works etc) have chemical contamination which results in aggressive ground conditions. Often the problems are unique to a particular site in terms of the type and level of contaminant, associated soil and water regimes and the materials being proposed (or having been used) for construction.

In aggressive ground conditions the target material is subject to gradual deterioration or decay, ranging in effect from leaking pipes and damaged cables to structural failures. (A list of the range of construction materials and aggressive contaminants can be found in the CIRIA Report 98.) Note that such substances or the products of the reaction between materials and contaminants can affect:

- Site workers engaged in investigation/construction
- Employees via contact with the materials or gaseous emissions
- Third parties adjacent to the site.

For each individual site there is a range of physical, chemical and biological factors which often act together to damage the fabric of buildings. Table 8.1 provides a simplified summary.

The extent and potential threat from aggressive ground conditions is directly related to the contaminative history of the site in question. Leaks, spillages, waste deposits or storage areas may all have contributed to the specific ground conditions present. The presence of made-ground or filled areas should always be treated with suspicion until the nature of the fill has been ascertained either from research or from a physical intrusive site investigation.

Chemical Agents	Physical Factors	Biological Agents
Acids	Ground Permeability	Fungi
Alkalis	Temperature (+ variation)	Algae
Salt Solutions	Oxygen concentration	Bacteria
Organic substances	Moisture	Insects (ants/termites etc)
Gases	Material Stress	Vermin (rodents or pigeons)
	Wetting/Drying Cycles	Invasive plants
	UV Light	

Table 8.1: Physical, chemical and biological factors
affecting fabric of buildings

Toxic or hazardous materials affecting people and the environment

There is an extensive group of materials or substances which directly through contact and/or exposure can cause harm to:

- Employees on site
- Site investigation workers
- Construction workers
- Maintenance staff.

The following is a limited description of some of these materials. Appendix 7 gives the ICRCL tentative trigger concentrations for a range of contaminants from ICRCL Guidance Note 59/83 (July 1987), *Assessment and redevelopment of contaminated land*. In addition, and for comparative purposes, the Dutch values for soil and groundwater contamination are also provided, as are the proposed Greater London Council Guidelines for Contaminated Soils which were offered at the Conference on Reclamation of Contaminated Land, Eastbourne, October 1980. Note that there is a definite move towards site-specific risk-based corrective action thresholds being set for individual contaminants, although a detailed description of this methodology is beyond the remit of this chapter. See the Environment Agency's Technical Series of Reports (1996) and the ASTM standards for further information (1995).

Metal compounds

Arsenic, cadmium, chromium, lead and mercury are all potentially hazardous metals; the hazard level ranges widely between the different metals and between different compounds of the same metal. For example, elemental cadmium (Cd) is a soft blue-white metal or grey/white powder. It is often a waste product of mining, smelting, pigment and paint manufacture, electroplating and the manufacture of batteries, alloys and solders. It is highly toxic when inhaled as the metal, or the metal oxide, in fumes or dust. Medically, this translates into chemical pneumonitis, hypertension, prostatic cancer, renal damage and failure. Thus in all its forms, including soluble cadmium compounds in water, it is a significant hazard (recommended ICRCL value 3 mg/kg in gardens – see Appendix 7).

Chromium is a hard, brittle, grey metal, but as a compound can appear in a variety of colours. It occurs naturally in the environment, within wastes from chrome-plating, anodising, chemical industries and pigment manufacture. Elemental chromium and chromium compounds in their trivalent state are relatively non-toxic. The ICRCL only recommends investigation where soil concentrations exceed 600 mg/kg for use in residential gardens. However, hexavalent chromium (such as chromium oxide, chromates and dichromates) are all soluble in water and easily absorbed in tissues. They cause irritating and corrosive effects on animal tissue, produce ulcers, respiratory tract irritation and ulceration of the nasal septum. This is reflected in the ICRCL tentative limit of 25 mg/kg for hexavalent chromium (see Appendix 7).

Acids and alkalis

Acids (with a pH of less than 7) are widely used in fertiliser and chemical industries, metal preparation and finishing, plastic manufacture, food and nylon industries. Alkalis (exceeding a pH of 7) are often the by-product of glass, chemical, paper and fertiliser industry manufacture. Both acids and alkalis can be corrosive to tissue, with damage occurring at, for example, 0.6% nitric acid, 1% sulphuric acid, 0.1% NaOH and 10% ammonia solution. They can cause respiratory damage, eye damage and even death in extreme cases. Furthermore, contact with other chemical substances can result in explosion or flammability. The principal hazards are to site investigators, redevelopment workers and site users.

Cyanides

Cyanide salts and solutions present a significant risk to humans and are particularly associated with plating works, heat treatment works, photographic processes, gasworks and pigment manufacture. They are absorbed into the body from all routes, reducing uptake of oxygen by tissues, and death if in excess of 50–100 mg are ingested. The most frequently encountered sources of cyanide contamination are old coal gas sites, of which there are usually several in each large city and at least one in each small town. Many of these were redeveloped when natural gas replaced town gas manufacture in the United Kingdom in the 1960s, although many of the remaining British Gas sites still have considerable problems. The cyanide was an impurity in the process which was removed using iron oxide, forming spent oxide. This was produced in large quantities and often landfilled on the old town gas properties or on adjacent lots. Along with the other contaminants it represents a significant health hazard.

Organic compounds

Coal tar
This is found on gas sites and is a viscous, black, highly combustible material. It can represent a hazard by inhalation (due to benzene/toluene) and a cancer risk (due to polycyclic aromatic hydrocarbons (PAH)). Skin/eye contact can cause severe irritation, burns and cancer.

Phenols
This is a residue (usually liquid) of town gas works and pharmaceutical, dye and oil-refining industries. It is toxic by inhalation, contact and ingestion, with tissue damage at 1% or more concentration: 10–30g can prove fatal if ingested.

PCBs
Polychlorinated biphenyls were widely used in transformers, capacitors, coolants, hydraulic oils and heat exchangers prior to 1975. Toxic and carcinogenic, these complex made substances do not degrade in the environment. The only known reliable and proven method of destruction is high temperature incineration.

Asbestos

Perhaps the most widely known hazard and one with which the property professional is almost certainly familiar, there are three types of asbestos: white (chrysoltile), brown (amosite) and blue (crocidolite). All are chemically inert, fibrous and heat resistant. They have been widely used throughout industry in a range of industrial appliances and materials, this use extending to domestic users. Asbestos is an irritant, but also a proven and effective carcinogen. Asbestos has no known safe level of exposure, although the hazard represented can generally be described as increasing in the order white → brown → blue, although this progression is very simplistic since there are many more types of asbestos (a naturally occurring mineral) which exist.

A brief review of some of the applications in which asbestos has been used, and may therefore be found by the auditor is revealing:

- Pipe insulation
- Boiler lagging
- Heating elements
- Wall insulation
- Acoustic panels
- Brake lining
- Ship building
- Carriage building (cars and trains)
- Putty
- Linoleum
- Roofing sheeting.

The risk from asbestos only becomes a hazard when people are exposed to the fibres, particularly in confined spaces. While asbestos can be found in virtually every industrial building over 30 years old, contact with the material by employees is often negligible. Only when damaged or removed during refurbishment and demolition is there a real risk. The danger for the purchaser or their funding institutions during acquisition is that redevelopment of a site may reveal asbestos. This will need specialist removal and disposal to a licensed landfill and may incur considerable and significant expense. On a recent audit of a large 150-acre industrial site in the south-east, the asbestos removal costs were estimated at £650,000.

8.3 **Sources of contamination**

Contamination of land can result from a number of causes such as:

- Storage and transport of materials
- Leaks and spillages
- Static emissions (particulates or volatile emissions)
- Waste disposal
- Premises alterations
- Sewerage sludge.

Hence, for example, a large spillage of petrol at a filling station could lead to the site and the surrounding area becoming contaminated with petrol and its associated vapours.

It is necessary for an auditor to be able to determine whether a site is potentially contaminated. If the auditor has grounds to believe that there is a potential risk of the site being contaminated, then further work can be undertaken to confirm this risk.

One of the prime means of determining the potential for a site to be contaminated is by examining its present and past uses, and determining whether any of these uses could have resulted in contamination. Certain uses (*e.g.* as a engineering works) are much more likely to result in contamination than others (*e.g.* an office block).

A list of potentially contaminative uses was published by the DoE in 1991 as an appendix to the "Registers of Land that may be Contaminated". This document was only ever issued in consultation form, but is now commonly used as a guide to determine whether a past or present use of the site has the potential to cause significant contamination. The document, which was introduced in EPA 1990, was firmly squashed by the influence of the powerful property lobby which was concerned by the potential effects it would have on the value of properties. A copy of the list of potentially contaminative uses in this register is included in Appendix 4.

The usefulness of this list is slowly being replaced by the publication of a number of "contaminated land profiles" published by the DETR (formally the DoE), which provide information about the nature of potentially contaminative industries and upon the specific contaminants that may be associated with them. These are:

- Airports
- Animal and animal products processing works

- Asbestos manufacturing works
- Ceramics, cement and asphalt manufacturing works
- Chemical works:
    coatings (paints and printing inks) manufacturing works
    cosmetics and toiletries manufacturing works
    disinfectants manufacturing works
    explosives, propellants and pyrotechnics manufacturing works
    fine chemical manufacturing works
    inorganic chemicals manufacturing works
    linoleum, vinyl and bitumen-based floor covering manufacturing works
    mastics, sealants, adhesives and roofing felt manufacturing works
    organic chemicals manufacturing works
    pesticides manufacturing works
    pharmaceuticals manufacturing works
    rubber processing works (including works manufacturing tyres or other rubber products)
    soap and detergent manufacturing works
- Dockyards and dockland
- Engineering works:
    aircraft manufacturing works
    electrical and electronic equipment manufacturing works (including works manufacturing equipment containing PCBs)
    mechanical engineering and ordnance works
    railway engineering works
    shipbuilding repair and ship breaking (including naval shipyards)
    vehicle manufacturing works
- Gas works, coke works and other coal carbonisation plants
- Metal manufacturing, refining and finishing works:
    electroplating and other metal finishing works
    iron and steelworks
    lead works
    non-ferrous metal works (excluding lead works)
    precious metal recovery works
- Oil refineries and bulk storage of crude oil and petroleum products
- Power stations (excluding nuclear power stations)

- Pulp and paper manufacturing works
- Railway land
- Road vehicle fuelling, service and repair:
    garages and filling stations
    transport and haulage centres
- Sewage works and sewage farms
- Textile works and dye works
- Timber products manufacturing works
- Timber treatment works
- Waste recycling, treatment and disposal sites:
    drum and tank cleaning and recycling plants
    hazardous waste treatment plants
    landfills and other waste treatment or waste disposal sites
    metal recycling sites
    solvent recovery works
- Profile of miscellaneous industries incorporating:
    Charcoal works
    Dry cleaners
    Fibreglass and fibreglass resins manufacturing works
    Glass manufacturing works
    Photographic processing industry
    Printing and bookbinding works

8.4 **Contaminated groundwater and surface water**

Groundwater is derived from a number of sources, but the majority of it comes from rainfall (water which has evaporated from the sea) and melting snow. This form of groundwater is correspondingly termed "meteoric groundwater". Infiltration is the passage of water through the surface of the ground, and its downward movement to the saturated zone is known as "percolation". Groundwater flow is the movement of water in the zone of saturated ground, where it returns to the sea via rivers, lakes and streams (the surface water system) to correspondingly evaporate and return to the land in the form of rain or snow. It can therefore be seen that a cycle of events exists: precipitation on land, infiltration and percolation, groundwater flow to surface water systems, return to the sea, evaporation followed by precipitation. This cycle of water is termed the "hydrological cycle".

Aquifers

Rocks and soils that transmit water with ease through their pores and fractures are called "aquifers". Those rocks and soils which do not transmit readily are called "aquicludes". Typical aquifers include gravel, sand, sandstone, chalk, limestone and fractured igneous and metamorphic rocks. Typical aquicludes are clay, mudstone, shale and unfractured igneous and metamorphic rocks. The majority of soils transfer water through their pores, whereas rocks tend to transfer the majority of their water through both pores and fractures. Although similar volumetric rates may be experienced in soils and rocks, the actual flow rates experienced in fractured rocks is usually higher and correspondingly the containment time is much shorter.

Aquifers vary in their general and hydraulic properties and are classified as fractured, fracture/intergranular and intergranular. These properties, particularly in the upper unsaturated zones, form the basis for the assessment of groundwater vulnerability, which in turn is mapped by the Environment Agency. The Agency subdivides permeable strata into two distinct types: highly permeable (major aquifers) and variably permeable (minor aquifers).

Minor aquifers (variable permeability) can be fractured or potentially fractured rocks, which do not have a high primary permeability including unconsolidated deposits. Although these aquifers will seldom produce large quantities of water for abstraction, they are important both for local supplies and in supplying the base flow for rivers.

A major aquifer is defined as a "highly permeable formation, usually with a known or probable presence of significant fracturing". Major aquifers are highly productive and can support a large number of abstractions for public supply and other purposes.

In addition, aquifers may be found within either particular strata or fractured zones. A confined aquifer is a stratiformed aquifer which is buried; correspondingly, an unconfined aquifer is one which is exposed (see Fig 8.1). The replenishment of most aquifers by infiltration (a process otherwise known as "recharge") occurs over the exposed section of the aquifer, and it is this unconfined portion of the aquifer where the greatest groundwater circulation occurs.

Geological boundaries define the volume of an aquifer, and hydrogeological boundaries (especially the water table) define the volume of the water stored within it. Common geological boundaries include stratification of aquifers, termination of aquifers by faults and aquifers positioned against aquicludes.

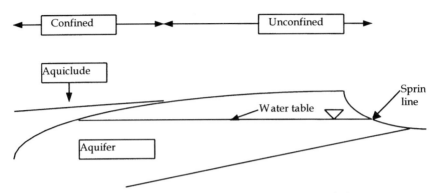

*Figure 8.1: Unconfined outcrop of scarpface with dip
slope confined beneath aquiclude.*

Groundwater Flow

Since a high proportion of rain falls on higher ground and the outlets of
groundwater systems are rivers, lakes and the sea, it can be seen that
groundwater flows tend to move through permeable rocks from high
ground to lower ground at a rate, which is determined by the aquifer's
permeability, and the hydraulic gradient, which is reflected by the
gradient of the water table.

A perched water table occurs where there is an impersistent zone of
saturation sitting on a limited impermeable strata, which is raised
above the main water table (see Fig 8.1).

Pollutant pathways

When dealing with pollution of the aquatic environment, sources of
pollution can be divided into two distinct categories: point sources and
distributed sources. A distributed source would commonly include
agricultural run-off (*e.g.* surface run-off and infiltration containing
fertilisers, pesticides etc), surface run-off from roads containing oils,
rubber and greases, and leachate from contaminated land. Point
sources of pollution are more commonly associated with industrial uses
and may include sudden releases resulting from failure/rupture of
pipelines, catastrophic failure of liquid storage tanks, accidents
involving road tankers or discharges in breach of discharge consents
into surface or groundwaters. Furthermore, long-term releases should
also be considered including leachate from landfills and contaminated

land, leaks in pipework, leaks associated with liquid storage tanks and losses from industrial processes.

Liquids are the main source of pollution into surface and groundwaters, although soluble solids and fine-grained solids suspended in the water may also be important. Contaminants can be divided into those which are soluble in water and those of low solubility. Soluble contaminants will migrate reflecting the movement of the groundwater/surface water flow. Low solubility contaminants, however, can be further subdivided into light non-aqueous phase liquids, which float on the water table as a layer moving in the direction of the slope of the water table, and dense non-aqueous phase liquids, which sink through the water table and have their behaviour controlled more by geological features rather than hydraulic features.

When a pollutant is released into the ambient environment, it does not necessarily enter the aquatic environment immediately, but may alternatively follow one of the following pathways:

(1) The pollutant may enter a secondary containment system, such as a bunded area, which may have an unsealed base allowing infiltration into the soil.
(2) The pollutant may enter the storm water drainage system. This system may enter water courses directly, or may discharge to the public sewer, or to a soak-away.
(3) The pollutant may enter an area which comprises of a superficial layer of artificial made ground. Made ground may include soil, brick or concrete rubble or demolition wastes. In addition, trenches around cables or pipework may have been backfilled. These backfilled trenches can be highly permeable and can offer migratory pathways for released pollutants.

In looking at the nature of the surface drainage system, and at the underlying geology of a site, it is possible to predict whether a pollution incident will impact surface waters, groundwaters or both. Note that even if there is a thin impermeable layer between a site and an aquifer, it would only require a small, backfilled trench to provide a downward migratory pathway for an potential pollutant.

By a similar manner, when undertaking a physical investigation of a site, it is important to ensure that any resulting boreholes do not provide a migratory pathway for pollutants. Boreholes should therefore be backfilled according to a rigorous specification, including staged compaction of the backfill material or the use of bentonite clay to seal the individual strata from each other during the drilling process.

Chapter 9

· **Site Remediation** ·

9.1 **Introduction**

The decision of whether remedial action is required on a site following a Phase II intrusive investigation will depend upon a number of factors:

(a) the nature and source of the contamination (*i.e.* whether it is discontinued or continuing, whether it involves chemical or organic substances, or solid, liquid or gaseous substances);
(b) the extent of contamination, namely its vertical limits and lateral boundaries;
(c) the consequences of the contamination; and
(d) the intended use of the site.

Before undertaking any remedial works on a site, it is important to identify the:

(a) vertical and lateral extent of soil and groundwater contamination;
(b) potential alternatives with regard to clean-up of the site;
(c) feasibility of implementing these alternatives; and
(d) selection and implementation of the most appropriate clean-up activities.

There are a number of alternatives with regard to remedial action to soil and groundwater contamination. In addition, with increases in technological advancement, the choice is becoming significantly greater as new techniques are accepted by industry. Remedial action techniques fall into three broad categories:

(1) *Removal.* Excavating the material and treating it either elsewhere on-site in controlled circumstances, or disposal off-site at a licensed tip.
(2) *Covering up.* Leaving the material in situ (*e.g.* containment by

horizontal or vertical sub-surface barriers and/or isolation of the site by covering it).

(3) *Cleaning up*. This may be off-site or on-site. It may include physical, biological or chemical methods or reducing or eliminating hazards associated with contaminants.

In the United Kingdom there is a preference for leaving contamination in place and dealing with it on-site. The arguments concerned with this form of treatment are concerned with keeping the danger to human health to a minimum. There is a strong school of thought that moving contaminated material from one place to another does not solve the problem, but merely transfers the problem elsewhere.

9.2 Commonly used methods for remediation – groundwater

Removal of floating product

The removal of floating product from groundwater can be achieved with skimmer pumps or passive collectors installed in wells, pits or trenches. A skimmer pump is designed to collect the liquids floating on top of the groundwater. The pumps are equipped with intakes that only allow the floating liquids to enter the pump storage chamber, and the float can move with the water table's natural fluctuations.

A passive collector is similar in design to a skimmer pump, except that it does not contain a mechanism to transfer the collected floating product to a holding tank. Instead, the collected product must be manually emptied when the collector becomes full. As with the skimmer, the passive collectors can move with the water table's natural fluctuations.

Since the floating product acts as a continuing source of groundwater contamination, as a proportion of the product will normally dissolve into the groundwater, any removal of floating product will lessen the potential for continued groundwater contamination, contaminant migration and toxic exposure.

Groundwater abstraction and on-site treatment

Groundwater abstraction and treatment involves pumping groundwater from wells located at a site and treating/removing the contaminants. Abstracted groundwater can be treated by oil/water separation, air stripping/aeration, ultraviolet (UV) oxidation, carbon absorption, electrocoagulation or bioremediation. Treated groundwater can be disposed of by re-injection on-site using infiltration galleries or injection wells, or discharged to the local wastewater treatment plant. This method has historically been used for restoring contaminated aquifers, although they are generally not effective enough to lower contaminant concentrations to meet drinking water criteria.

Air sparging (in situ air stripping)

Air sparging is a mass transfer process that forces air below the water table and uses the bubbles generated to mobilise the dissolved contaminants upwards. Factors that affect the success of air sparging include volatility, solubility and biodegradability of the contaminants, permeability of the soils in both the saturated and unsaturated zones, depth of the groundwater and the site geology. Air sparging is effective at reducing volatile components. In general if the contaminants have a Henry's law constant of 10^{-5} atm-m^3/mole or greater they are suitable for removal by this method. (Henry's law constant is the ratio of the concentration of the substance in air divided by its concentration in water. Most Henry's law constants are approximated as the ratio of the vapour pressure of a substance divided by the water solubility.)

An additional benefit of air sparging is its ability to increase the biodegradation of contaminants in the groundwater and soils by increasing the oxygen content through the bubbling action. By increasing the oxygen content, the increased aerobic biological activity can metabolise (*i.e.* break down most of the contaminants present).

In situ bioremediation

Bioremediation techniques for groundwater are very similar to those described for soil (see below). The only significant difference is that the oxygen and nutrients are injected below the water table rather than above it.

Electrocoagulation

Many wastewater contaminants are held in solution owing primarily to electrical charges. Bacteria, algae, oils, clays, carbon black, silica, phosphate, nickel, lead, chromate and other ions are only some examples. Neutralisation of these charges and the subsequent precipitation of these contaminants can be achieved either by chemical or by electrochemical means. Electrocoagulation systems have been employed for years in the treatment of wastewater. Most rely upon high voltages to produce a strong electromagnetic field to disrupt the attraction of the particles, allowing suspended contaminants to precipitate.

In the past, these systems have shown good contaminant removal compared to chemical precipitation; nevertheless, higher capital and operation costs, along with lower flow rates, have reduced the use of these systems. In today's environment, chemical addition is becoming less acceptable because of more stringent regulations. Solid residues are classified as hazardous and treatment levels are more difficult to achieve. Lower operating costs, higher flow rates and better knowledge of the process have moved electrocoagulation to the forefront of water treatment technologies.

9.3 Commonly used methods of remediation – soil

Excavation and off-site disposal

A standard option for the remediation of contaminated soils is excavation and off-site disposal. Contaminated soils are excavated from the site using standard mechanical excavation equipment and are transported to a waste disposal facility, usually a landfill. Groundwater which infiltrates into the open excavations is pumped out and collected in tankers for off-site disposal. Excavated soils are normally replaced with clean fill. The removal of the most heavily contaminated soils will reduce the mass of contamination that could potentially infiltrate into the underlying groundwater and migrate off-site, or affect humans exposed to the soils.

Vapour extraction

This method is effective in removing volatile contaminants from soils by volatilisation and vapour transport. Soil vapour extraction uses extraction points similar in design to groundwater wells, which are installed in the unsaturated zone above the water table. The points are connected to a vacuum source, which draws the volatile contaminants from the soils into the points. The movement of air through the pore spaces of the soils mobilises the contaminants through volatilisation.

A soil vapour extraction system can remove residual contamination located in unsaturated soils above the water table, thus eliminating ongoing sources of groundwater contamination. Soil vapour extraction is capable of treating soils beneath structures and equipment. The depth of groundwater will affect the depth of soil that can be treated. Compounds which can be removed from the soil using soil vapour extraction might potentially require treatment prior to discharge to the atmosphere. Treatment of the vapour is generally performed using carbon absorption, thermal incineration, catalytic oxidation or bioremediation.

In situ bioremediation

Bioremediation provides an environment that enables the indigenous microbes to metabolise the contaminants present in the soils at the facility. The manipulation of the soil environment is achieved through the addition of nutrients, oxygen and water. Some of the methods of providing these additions include horizontal or vertical infiltration points, surface applications and in situ mixing.

Bioremediation is capable of degrading organic compounds into more basic degradation products, which are generally less harmful to the environment. The aerobic bioremediation process produces carbon dioxide and water vapour as a result of the microbial degradation. Bioremediation will successfully reduce the toxicity and concentration of contaminants, but will not affect their mobility.

Chapter 10

Developments in
Environmental Auditing

10.1 **Introduction**

International standards on environmental auditing are now been published, as are standards of on-site assessment, investigation and intrusive investigations. The standards being developed are intended to guide organisations, auditors and their clients on the general principles common to the conduct of environmental audits. The standards provide definitions on environmental auditing, the related terms and the general principles of environmental auditing. To date there are three international standards on auditing:

EN ISO 14010:1996	Guidelines on Environmental Auditing – General Principles
EN ISO 14011:1996	Guidelines for Environmental Auditing – Audit Procedures – Auditing of Environmental Management Systems.
EN ISO 14012:1996	Guidelines for Environmental Auditing – Qualification Criteria for Environmental Auditors.

Each of these is discussed below. An international standard on auditing (ISO 14015) is expected to be available for comment in draft form in mid-2000. No information is available at present on the content of 14015, although further information is available from the British Standards Institute (see Useful Contacts, p 169) .

10.2 **EN ISO 14010:1996**

This international standard defines an environmental audit as a systematic, documented verification process of objectively obtaining and evaluating audit evidence to determine whether specified environmental activities, events, conditions, management systems, or information about these matters conform with audit criteria, and communicating the results of this process to the client. The standards provide a number of definitions for the various terms associated with environmental auditing.

This standard states that an environmental audit should focus on clearly defined and documented subject-matter, and should only be undertaken if there is:

- Sufficient and appropriate information about the audit subject-matter
- Adequate resources to support the audit process
- Adequate co-operation from the auditee.

The audit scope (see p 59 above), which is determined by consultation with the client and the lead auditor, should be designed to meet the audit objectives, which are established by the client.

Since objectivity is one of the key elements of an environmental audit, it is essential that the audit team members are independent of the audit activities. Therefore, if there are internal audit team members from the organisation being audited, they should not be accountable to those directly responsible for the subject-matter being audited. In addition, unless otherwise stated by law, the audit findings should be confidential and consequently the audit team members should not disclose information or documents to third parties without the client's express permission.

One of the distinctive features of environmental auditing is its systematic approach and methodology. The standard states that for all types of environmental audit, the methodologies and procedures should be consistent. Therefore, the procedures for one type of environmental audit should only differ from those of another where it is essential to the specific character of that environmental audit.

The collection, assessment and quality of audit information is essential to the successful completion of an environmental audit. The information which is collected during an environmental audit should be interpreted and documented, and presented as evidence during the reporting stage of the process. Correspondingly, the audit evidence should be of such a quality and quantity that competent environmental auditors working independently of each other will reach similar audit findings upon evaluating the same evidence.

The standards highlight the fact that the reliability of an

environmental audit is limited since the audit is undertaken during a limited time period, and therefore the information available is only a proportion of the total available. Consequently, the auditor should balance the audit requirements with any inherent limitations. This balance should then be accounted for adequately during the planning and conduct of the audit.

10.3 **EN ISO 14011:1996**

This is the first of the standards which deals specifically with undertaking a certain form of audit. The standard establishes the audit procedures for planning and conducting an audit of an environmental management system (EMS) to determine conformity with EMS audit criteria.

The standard defines an EMS as that part of the overall management system that includes organisational structure, planning activities, responsibilities, practices, procedures, processes and resources for developing, implementing, achieving, reviewing and maintaining the environmental policy (ISO 14001). As with any environmental audit, there should be specific objectives in undertaking the audit. Typical objectives of an EMS audit are given as to:

(a) determine conformity of an auditee's EMS with the EMS audit criteria;
(b) determine whether the auditee's EMS has been properly implemented and maintained;
(c) identify areas of potential improvement in the auditee's EMS;
(d) assess the ability of the internal management review process to ensure the continuing suitability and effectiveness of the EMS; or
(e) evaluate the EMS of an organisation where there is a desire to establish a contractual relationship, such as with a potential supplier of a joint venture partner.

The standard sets out the roles, responsibilities and activities undertaken by the audit team, lead auditor, auditee and auditors during the auditing process, and as would be expected generally follow the guidelines laid down in ISO 14010.

10.4 **EN ISO 14012:1996**

This standard provides guidance on the qualification criteria for environmental auditors and lead auditors. The standard states that an auditor should have completed at least secondary education or equivalent. In addition, auditors should have appropriate work experience which contributes to the development of skill and understanding in some or all of the following:

- Environmental science and technology
- Technical and environmental aspects of facility operations
- Relevant requirements of environmental laws, regulation and related documents
- Environmental management systems and standards against which audits may be conducted
- Audit procedures, processes and techniques.

According to the standards, if an auditor has only completed secondary education, then he should also hold five years of appropriate work experience, although this criteria can be reduced by the successful completion of post-secondary full- or part-time education covering one or all of those subjects listed above. If an auditor has obtained a degree, then he requires four years of relevant work experience, with post-graduate experience reducing this criterion if covering one or all of the subject areas listed above.

In addition to the formal requirements already mentioned, environmental auditors will also be expected to have undertaken both formal and on-the-job training to develop competency in carrying out environmental auditing, although if competence can be demonstrated from accredited examinations then the requirement for formal training may be waived. Formal training is specified within the standard as addressing the five subjects listed above. On-the-job training is described in the standard as constituting a total of 20 equivalent workdays of environmental auditing, including involvement in the complete audit process.

For an auditor to qualify as a lead auditor, the standard states that in addition to the personnel skills necessary to ensure efficient management and leadership of the audit process, the auditor will be required to meet the following requirements:

(1) Either participation in the entire audit process for a total of 15 equivalent workdays of environmental auditing, for a minimum of three additional complete environmental audits and participation as acting lead auditor, under the supervision and guidance of a lead auditor, for at least three of these audits; or

(2) demonstration of those attributes and skills to the audit programme management or others, by means such as interviews, observation, references and/or assessments of environmental auditing performance made under quality assurance programmes.

· Glossary of Terms ·

This Glossary is based on the combined glossaries of the Environmental Compliance Manual and the Basic Guide for Environmental Inspection, published by the Environmental Assessment Association, Scottsdale, Arizona, US.

Term	Definition
Abatement	Controlling, reducing or eliminating existing pollution. Both a legislative and technical term.
Activated carbon	A highly absorbent form of carbon, used to remove odours and toxic substances from gaseous emissions or liquid effluent.
Ambient pollution	Pollution in the surrounding environment– sometimes used to describe background levels.
Aquifer	Geological formation, group of formations or part of a ground formation which is usually gravel or porous (*i.e.* capable of yielding water to wells or springs).
Bioaccumulation	Build up of toxins in fish and animal life, ultimately passed to humans or higher predator through the food chain.
Biodegradation	Process of decomposing. Primary biodegradation is when the product loses some characteristics. Ultimate biodegradation is when the product breaks down into its various elements.
Biological hazardous wastes	Any substance of a human or animal origin (other than food wastes) which is to be disposed of and could harbour or transmit pathogenic organisms including, but not limited to, pathological specimens such as tissues, blood elements, excreta, secretions, bandages and related substances.
Biological Oxygen Demand (BOD)	Measure of the biodegradable organic pollution present in a water course. (Measured mg/l). A high level of organic matter (with high BOD) in effluent

stimulates microbial growth which in turn removes oxygen from water necessary to sustain aquatic environment.

By-product	A material produced without separate commercial intent during the manufacture or processing of other materials or mixtures.
Carcinogenic	A substance or preparation which, if inhaled or ingested or penetrates the skin, may induce cancer.
Carriage list	Lists those hazardous substances approved by the DETR regulations.
Chemical Oxygen Demand (COD)	Measure of the amount of potassium dichromate needed to oxidise (organic and inorganic) reducing material in a water sample. Generally more useful (and produces higher readings) than BOD as materials such as cellulose react with dichromate but not with oxygen.
Combustible liquid	Any liquid having a flash point above 100° Fahrenheit as determined by tests.
Combustion	Chemical reaction whereby fuel combines with oxygen.
Commercial waste	All solid waste emanating from establishments engaged in business. This category includes, but is not limited to, solid waste originating in stores, markets, office buildings, restaurants, shopping centres and theatres.
Compliance monitoring programme	A programme used to determine whether ground-water performance standards are exceeded.
Condensation	Reaction by which a heavier substance is produced than the original (*e.g.* water in air turning into liquid).
Container	Any portable device in which a material is stored, transported, treated, disposed or otherwise handled.
Contingency plan	A document setting forth an organised, planned and co-ordinated course of action to be followed in order to prevent pollution in case of fire, explosion or discharge of hazardous waste constituents which could threaten human health and the environment.
Controlled water	All fresh and saline natural waters up to the

UK offshore territorial limit, including rivers, streams, lochs, estuaries, coastal waters and groundwater. Legislation relating to the discharge of effluent (other than to sewer) applies to controlled water. The statutory definition of "controlled waters" is given in the Water Resources Act 1991, s 104(1) and the Control of Pollution Act 1974, s 30A(d).

Cradle-to-grave
The tracking of the source, quantity, concentration and type of hazardous waste from generation through to final disposal.

Decomposition
Breakdown of a material or substance (by heat, chemical reaction, electrolysis, decay or other processes) into parts or elements or simpler compounds.

Decontamination
The process of making any person, object or area safe by absorbing, destroying, neutralising or making harmless by removing biological or chemical agents.

Discharge consent
Permit/authorisation issued by the sewerage undertaker (England & Wales) or water authority (Scotland) allowing discharge of trade effluent to public sewer.

Discharge/hazardous waste discharge
The accidental or intentional spilling, leaking, pumping, pouring, emitting, emptying or dumping of hazardous waste into or on any land or water.

Disposal facility
A collection of equipment and associated land area which serves to receive waste and dispose of it. The facility may have available one, many or all of the large number of disposal methods.

Domestic waste(s)
Solid waste, garbage and rubbish which originate in residential areas.

Dump
A land site at which waste is disposed of in manner which does not protect the environment, is susceptible to open burning, or is exposed to the elements, vermin and/or scavengers.

Duty of care
A duty to prevent the escape of waste (under EPA 1990, s 34) placed on persons holding (*e.g.* importing, treating or disposing) controlled waste in a commercial capacity. Persons to whom waste is transferred must be authorised and the consignments accompanied by proper documentation.

EC Black List   List of substances (set out in Directive 76/464/ECC) considered particularly harmful to the water environment (*e.g.* cadmium and organophosphates) and subject to special controls *e.g.* licensed emission limits). Some of these are UK Red List substances.

EC Grey List   Substances considered harmful (but, less so than Black List substances) when discharged to water, although still considered to have "a deleterious effect on the aquatic environment".

Ecotoxicological   A substance/preparation, which presents/may present immediate/delayed risks to the environment.

Effluent   (1) Solid, liquid or gas wastes which enter the environment as a by-product of man-oriented processes. (2) The discharge or outflow of water from ground or sub-surface storage.

Encapsulation   The complete enclosure of a waste in another material in such a way as to isolate it from external effects such as water or air.

Explosive   Any chemical compound, mixture or device, the primary or common purpose of which is to function by explosion (*i.e.* with substantially instantaneous release of gas and heat).

Extraction   Process for dissolving one part of a mixture by means of a liquid solvent acting on one property only (*e.g.* grease may be removed from fabric using petrol).

Feasibility study   A detailed examination of the technical, environmental, engineering, economic, legal and practical suitability of a proposed facility or technology for use at a specific location.

Final closure   The measures which must be taken by a facility to render the landfill portion environmentally innocuous when it determines that it will no longer accept waste for treatment, storage, or disposal on the entire facility.

Flood plain   The lowland that borders a river, which is usually dry, but is subject to flooding when the river overflows its banks.

Fugitive emissions   Emissions produced by process but not released through chimney or stack emissions (*e.g.* gas leakages from pipes).

Grabsample	A single sample of wastewater taken at neither set time nor flow.
Green List	Substances classified under EC Regulation 259/93 as normally non-hazardous which may be shipped for recovery without attracting full regulatory controls.
Greenhouse gases	Gases which trap the sun's rays inside the earth's atmosphere causing the temperature to rise (*e.g.* carbon dioxide, nitrous oxide methane, perofluorocarbons, chlorofluorocarbons and sulphuric fluoride).
Ground-level ozone	Ozone produced at ground-level by hydrocarbons and nitrogen oxides combining in sunlight. Can cause eye, nose, throat and lung irritation.
Hazardous materials	In a broad sense, any substance or mixture of substances having properties capable of producing adverse effects on the health or safety of a human being.
Hazardous waste	A waste, or combination of wastes, which because of its quantity, concentration, toxicity, corrosiveness, mutagenicity or flammability, or physical, chemical or infectious characteristics may (1) cause, or significantly contribute to an increase in mortality or an increase in serious irreversible, or incapacitating reversible illness; or (2) pose a substantial present or potential hazard to human health or the environment when improperly treated, stored, transported or disposed of, or otherwise managed.
Hazardous waste landfill	An excavated or engineered area on which hazardous waste is deposited and covered. Proper protection of the environment from the materials to be deposited in such a landfill requires careful site selection, good design, proper operation, leachate collection and treatment and thorough final closure.
Hazardous waste site	A location where hazardous wastes are stored, treated, incinerated, or otherwise disposed of.
Heavy metals	High-density metallic elements (mercury chromium, cadmium, arsenic and lead) which are generally toxic to plant and animal life in low concentrations.

Hydrolysis	Process using water to break down substances. Hydrolysis has been suggested as a refuse disposal technique.
Inactive or inert waste	Waste which does not biodegrade (usually mineral material, concrete, bricks).
Incineration	An engineered process using controlled flame combustion to thermally degrade waste materials. Devices normally used for incineration include rotary kilns, fluidised beds and liquid injectors. Incinerators must meet clean air standards.
Industrial wastes	Unwanted materials produced in or eliminated from an industrial operation. They may be categorised under a variety of headings, such as liquid wastes, sludge wastes and solid wastes. Hazardous wastes contain substances which, in low concentrations, are dangerous to life (especially human) for reasons of toxicity, corrosiveness, mutagenicity and flammability.
Landfills	A conventional sanitary landfill is "a land disposal site employing an engineered method of disposing of solid wastes on land in a manner that minimises environmental hazards by spreading solid wastes in thin layers, compacting the wastes to the smallest practical volume and applying cover materials at the end of each operating day". A secure chemical waste landfill should be designed to provide complete protection to the quality of surface and sub-surface waters, thereby making site selection very important. Potentially hazardous wastes frequently require various types of pretreatment before they are disposed (*i.e.* neutralisation, chemical fixation or encapsulation). Some types of wastes also require segregated disposal.
Mitigation	The process of removing or eliminating the environmental problem.
Monitoring well	A well used to obtain water samples for water quality analysis or to measure groundwater levels.
Oxidation	Addition of oxygen to a compound.
Oxidising	Produces oxygen in a chemical reaction. Oxidising agents include ozone and nitrogen dioxide.

Oxidising agents	A chemical which gives up oxygen in chemical reactions or supplies an equivalent element (chlorine) to combine with a reducing agent. Also means the removal of hydrogen from a substance. Ozone and nitrogen dioxide are oxidising agents.
Ozone depleters	Substances which destroy the layer of ozone in the stratosphere which absorbs large amounts of harmful solar radiation in the form of ultraviolet light (*e.g.* HFCs, CFCs).
Particulate matter	A solid suspended in the air column.
Particulates	A collective air pollution term for grit, dust, fume, aerosol, smoke, etc. Particulates are fine solid or liquid droplets suspended in air; they can also carry *e.g.* smoke particles and sulphur dioxide which compounds the polluting effects.
Pesticides	A substance or preparation used for destroying any pest or protecting plants from harmful organisms.
Phosphorus	Non-metallic element occurring abundantly in minerals and all living matter. Manufactured by burning sand and carbon in an electric furnace.
Point Source	A discernible, confined and discrete conveyance, including, but not limited to, a pipe, ditch, channel, tunnel, conduit, well, discrete fissure, container, rolling stock, concentrated animal-feeding operation, or vessel or other floating craft from which pollutants are or may be discharged. Does not include return flow from irrigated agriculture.
Pollution	Contamination of air, water, land or other natural resources that will or are likely to create a public nuisance or to render such air, water, land or other natural resources harmful, detrimental or injurious to public health, safety or welfare, or to domestic, municipal, commercial, industrial, agricultural, recreational or other legitimate beneficial uses, or to livestock, wild animals, birds, fish or other life.
Polychlorinated biphenyls (PCB's)	A series of hazardous compounds used for a number of industrial purposes now found throughout the natural environment. PCBs are toxic to some marine life at concentrations of a few

ppb (parts per billion) and are known to cause skin diseases, digestive disturbances and even death in humans at higher concentrations. PCBs are persistent in the environment and do not easily decompose and biomagnify up the food chain.

Prescribed process   Processes which use and/or generate substances which are harmful if released into the environment (*i.e.* prescribed substances) and their emissions are controlled under EPA 1990.

Prescribed substances   Substances, defined under EPA 1990, considered to be harmful if released into the environment. Includes asbestos, glass fibre, halogens, organic compounds, oxides of carbon, nitrogen and sulphur, particulate matter and phosphorous.

Recycling   As commonly used, using discarded materials and objects in original or changed form rather than wasting them. Precisely used, refers to sending a material back into the process by which it was first formed.

Resource conservation   Reduction of the amounts of waste that are generated, reduction of overall consumption and utilisation of recovered resources.

Red List   Substances classified under EEC/259/93 as extremely hazardous with correspondingly strict controls on their movements.

Rubbish   Solid wastes which are not liable to rot, consisting of both combustible and non-combustible wastes, including paper, wrappings, cardboard, tin cans, garden clippings, wood, glass, bedding, crockery and similar materials.

Site   The property on which a facility is located. Two or more pieces of property which are divided only by public or private right(s) of way and which are otherwise geographically contiguous are considered a single site.

Solid waste   Any garbage, refuse, sludge from a waste treatment plant, water supply treatment plant, or air pollution control facility and other discarded material, including solid, liquid, semi-solid or contained gaseous material resulting from industrial, commercial, mining and agricultural operations, and from community activities. It does not include solid or dissolved material in domestic sewage, or solid or dissolved materials in irrigation return, flows or industrial discharges which are point sources.

Solvent
Liquid that is capable of dissolving another substance; used in a number of manu-facturing/industrial processes including the manufacture of paints and coatings for industrial and household purposes, equipment clean-up and surface degreasing in metal-fabricating industries.

Statutory nuisance
Nuisance (irrespective of common law) defined by legislation (*e.g.* EPA 1990, s 79(1)) which requires local authorities to inspect their area for nuisance such as smoke, fume, or dust.

Suspended particulates
Solid material trapped in air or water.

Total dissolved solids
Measure of solid material in water or effluent sampled after evaporation.

Total suspended solids
Measure of particulate matter in water or effluent.

Toxicological
Preparation which is an organic poison.

Trade effluent
Liquid (wholly or partly) produced in the course of any industry at trade premises with the exception of sewage. The statutory description is given in the Water Industry Act 1991, s 141 and the Sewerage (Scotland) Act 1968, s 59(1).

Transboundary pollution
Pollution which is produced in one region (or country) and carried across into another (*e.g.* by rivers or air masses).

Turbidity
Reduced visibility in atmosphere (other than clouds) or in liquid caused by presence particulates.

UK Red List
List of 23 high dangerous substances (derived from the EC Black List, plus asbestos) subject to strict regulatory control for release to water through trade effluent regulations.

Volatile Organic Compounds (VOCs)
Organic compounds (*e.g.* benzene, ethylene, propylene, acetone, but not methane) which produce photochemical oxidants by reaction with nitrogen oxides in the presence of sunlight. Emissions regulated as they contribute to both the greenhouse effect and depletion of the ozone layer.

· Bibliography ·

ADL, *Benefits to Industry of Environmental Auditing*, Centre of Environmental Insurance, Arthur D Little Inc, Cambridge, MA, US, 1983

Bamping, NC, *Insurers Discussion Group Environment Act 1995: Legal issues – an overview*, Partner, Barlow Lyde & Gilbert Solicitors (unpublished paper), 1997

Boulding, JR, *Practical Handbook of Soil, Vadose Zone and Groundwater Contamination. Assessment, Prevention and Remediation*, Lewis Publishers, 1995

Bowman, V, *Professional Environmental Management and Auditing*, 2nd ed, Cahners Publishing Co, 1992

Calabrese, E, and Kostecki, P, *Soils Contaminated by Petroleum: Environmental and Public Health Effects*, School of Health Sciences Division of Public Health, University of Massachusetts, US, 1988

CIRIA Report 98, *Material Durability in Aggressive Ground*, 1983

Coles, T, *Preparatory Review of Auditing*, Institute of Environmental Assessment, 1992.

Croner's Environmental Policy and Procedures, Croner Publications Ltd, 1996

Croner's Substances Hazardous to the Environment, Croner Publications Ltd, 1997

Gogen, R, *Environmental Audits in Connection with Property Purchases and Sales*, Presentation at the 78th Annual Meeting of the Air Pollution Control Association, 1985

Fogleman, V, *Insurers Discussion Group Environment Act 1995: Environmental Liabilities*, solicitor and member of Environmental Liability Group, Barlow Lyde & Gilbert Solicitors (unpublished), undated

Hay, G, and Morrell, J, *The Environment Act 1995: Implications for surveyors and valuers. An outline of the key points for discussion*, seminar held at the Strand Hotel, London, 1997

Ladd Greeno *et al*, *Environmental Auditing – Fundementals and Techniques*, 2nd ed, Centre for Environmental Assurance, Arthur D Little Inc, Cambridge, MA, US, 1988

Ledgerwood, G, *et al*, *The Environmental Audit and Business Strategy: A total quality approach*, Financial Times/Pitman Publishing, 1992

Newton, J, *Environmental Auditing*, 3rd ed, Cahners Publishing Co

Pierzynski, G, *et al*, *Soil and Environmental Quality*, Lewis Publishers, 1994

Side, J, *Notes on environmental policy, control and monitoring*, Heriot Watt University, unpublished, 1993

Smith, S, *Monitoring and Remediation Wells, Problem Prevention, Maintenance and Rehabilitation*, Lewis Publishers, 1995

Taylor, C, "Purchaser Protection Residential Property and the Contaminated Land Problem" (1999) 7(3) *Environmental Assessment* (the magazine of IEMA Ltd)

Wilson, LG, *et al*, *Handbook of Vadose Zone Characterization and Monitoring*, Geraghty and Miller Environmental Science and Engineering Series, 1995

Wilson, N, *Soil and Groundwater Sampling*, Lewis Publishers, 1995

Official publications

ASTM, *Risk-based Corrective Action Applied at Petroleum Release Sites*, American Society for Testing and Materials, Standard Guide E1739-95, 1995

ASTM, *Environmental Site Assessment for Commercial Real Estate*, American Society for Testing and Materials, Standard Guide E1527-94, "Standard Practice for Environmental Site Assessment Phase I Environmental Site Assessment Process"; E1528-93, "Standard Practice for Environmental Site Assessment Transaction Screen Process", 2nd ed, 1994

British Standards Institute, *Draft Code of Practice for investigation of Potentially Contaminated Sites*, 1998

CBI, Confederation of British Industry, Centre Point, 103 New Oxford Street, London, WC1A 1DU, ISBN 085201371X, 1990

CIA, *Guidance on Safety, Occupational Health and Environmental Protection Auditing*, Chemical Industries Association, 1991

DD175, *Code of Practice for the identification of potentially contaminated land and its investigation* (Draft for Development), Draft BSI Standard, 1988

DETR, *Fourth Consultation Paper on the Implementation of the IPPC Directive*, Department of the Environment, Transport and the Regions in partnership with the Welsh Assembly, August 1999

DETR, *Contaminated Land Implementation of Part IIA of the Environmental Protection Act*, Draft DETR Circular, 1999

DoE, Contaminated Land Research Report No 1, *A framework for assessing the impact of contaminated land and groundwater and surface water*, 1994

DoE, Contaminated Land Research Report No 2, Vol 1, *Contaminated Land Research Report Guidance on Preliminary Site Inspection of Contaminated Land*, prepared by Aspinwall & Co, 1994

DoE, Contaminated Land Research Report No 3, *Contaminated Land Research Report, Documentary Research on Industrial Sites*, 1994

EEP, *The Environmental Handbook* Series (5 vols), Executive Enterprises Publications Co, New York, 1988

Environment Agency, *A Methodology to Derive Groundwater Clean-up Standards*, R&D Technical Report P12, prepared by WRc, 1996

Environment Agency, *A methodology to determine the degree of soils clean-up required to protect water resources*, R&D Technical Report P13, prepared by Dames and Moore, 1996

ENDS Report No 236, September 1994.

ICC, *Environmental Auditing*, International Chamber of Commerce, Paris, 1989

ICC, *Effective Environmental Auditing*, International Chamber of Commerce, Paris, 1991

ICRCL, Guidance Note 59/83, *Assessment and redevelopment of contaminated water*, 2nd ed, Interdepartmental Committee on the Redevelopment of Contaminated Land, 1987.

UNEP/IEP, United Nations Environment Programme, Industry and Environment Office, Technical Report Series No 2, Environmental Auditing Report of a UNEP/IEO workshop, Paris (January 1989), 1990

Useful Contact Numbers and Information

This list of contacts is based on the lists supplied in Appendix D of the *Environmental Compliance Manual** and information held at EAL.

Environmental Auditors Registration Association (EARA)
Welton House
Limekiln Way
Lincoln LN2 4US
Tel: (01522) 540069
Fax: (01522) 540090

Environmental Data Association (EDA)
Unit 2
Redhouse Farm
Newtimber
West Sussex BN6 9BS
Tel: (01273) 857500
Fax: (01273) 857550

British Geological Survey
Durham Centre
Keyworth
Nottingham NG12 5GG
Tel: (0115) 936 3241
Fax: (0115) 936 3488

DETR Publications
Tel: (020) 8691 9191

British Standards Institute
Customer Services (020) 8996 9000
Fax: (01273) 857550
Fax: (0115) 936 3488

Stationery Office (formerly HMSO)
Tel (enquiries): (020) 7873 0011
Tel (orders): (020) 7873 9090
Fax (orders): (020) 7873 8200

* Published by Information for Industry Ltd, 4 Valentine Place, London, SE1 8RB, www.ifi.co.uk.

Regulatory contacts

Environment Agency

Enquiry line: (0645) 333111 (connects directly to your local office)
Pollution reporting: Tel: (0800) 807060;
Web: www.environment agency.gov uk;
Email: enquiries@environment-agency.gov.uk;
REPACs: telephone your regional HQ

Bristol HQ
Tel: 01454 624400
Fax: 01454 624409

North East region
HQ
Tel: 01454 624400
Fax: 01454 624409
North east region
HQ
Tel: (0113) 244 0191
Fax: (0113) 246 1889
Environment protection – Cyril McQuillian
IPC – Bob Barker
Waste – Offord Slater
Water quality – Richard Brook

Northumbria area
Tel: (0191) 203 4000
Fax: (0191) 203 4004
Environmental Planning Manager – John Burns

Dales area
Tel: (01904) 692296
Fax: (01904) 693748

Environmental Planning Manager – Don Ridley
Environmental Protection Managers Bob Pailor, Mike McNutty
Ridings area
Tel: (0113) 244 0191
Fax: (0113) 231 2116
Environmental Planning Manager – Darryl Toothill
Environmental Protection Managers Jan Davie, Gerald Morris, John Housham

Anglian region

HQ
Tel: (01733) 371811
Fax: (01733) 231840
Pollution Prevention & Control – Michael Pearson
IPC – Innes Garden
Waste – Peter Proctor
Water – Tony Warn

Northern area
Tel: (01522) 513100
Fax: (01522) 512927
Environmental Planning Manager – John Sweeney
Environmental Protection Managers – David Hawley & Matthew Clark

Central area
Tel: (01480) 414581
Fax: (01480) 413381
Environmental Planning Manager – Paul Waldron
Environmental Protection Managers David Hawley, Matthew Clarke

Eastern area
Tel: (01473) 727712
Fax: (01473) 724205
Environmental Planning Manager
Paul Hayward
Environmental Protection Managers
Geoff Philips, Alison Bramwell

Thames region
HQ
Tel: (01734) 535000
Fax: (01734) 500388
IPC – David Frith
Pollution & Prevention Control
Simon Read
Waste – Mike Fletcher
Water – Ian Adams

West area
Tel: (01734) 535000
Fax: (01734) 535900
Environmental Planning Manager
John Weir
Environmental Protection Managers
David Keeling, Barry Sheppard

North-east area
Tel: (01992) 635566
Fax: (01992) 645468
Environmental Planning Manager –
Malcolm Allen
Environmental Protection Manager
Bill Harris

South-east area
Tel: (01932) 789833
Fax: (01932) 786463
Environmental Planning Manager –
Paul Hudson
Environmental Protection Managers
Doug Greeves, Pete Lloyd

Southern region

HQ
Tel: (01903) 832000
Fax: (01903) 821832
IPC – Chris McDonald
Pollution & Prevention Control –
Arthur Tingley
Waste – Michael Baker
Water – Robert Edmunds

Hampshire area
Tel: (01962) 713267
Fax: (01962) 841573
Environmental Planning Manager –
John Adams
Environmental Protection Manager
Peter Kelly

Sussex area
Tel: (01903) 215835
Fax: (01903) 215884
Environmental Planning Manager –
Dave Watson
Environmental Protection Manager
Richard Hammond

Kent area
Tel: (01732) 875587
Fax: (01732) 875057
Environmental Planning Manager –
Colin Buckle
Environmental Protection Manager
Harvey Bradshaw

Isle of Wight
Tel: (01983) 822986
Fax: (01983) 822985
Contacts: see Hampshire

South-West region

HQ
Tel: (01392) 444000
Fax: (01392) 444238
IPC – John Hescott

Pollution & Prevention Control –
Martin Booth
Waste – Ian White
Water – Bill Grigg
Cornwall area
Tel: (01208) 78301
Fax: (01208) 78321
Environmental Planning Manager –
Judy Proctor
Environmental Protection Manager –
Brian Sinkins

Devon area
Tel: (01392) 444000
Fax: (01392) 444238
Environmental Planning Manager –
Malcolm Newton
Environmental Protection Managers
– Malcolm Chudley, Heather Baker

North Wessex area
Tel: (01278) 457333
Fax: (01278) 452985
Environmental Planning Manager –
Steve Chandler
Environmental Protection Managers
Andy Gardiner, Ian Legge

South Wessex area
Tel: (01258) 456080
Fax: (01258) 455998
Environmental Planning Manager –
Ron Curtis
Environmental Protection Managers
Bob Huggins, Howard Davidson

Midlands region
HQ
Tel: (0121) 711 2324
Fax: (0121) 711 5824
IPC – Robin Gaulton
Pollution & Prevention Control –
Andrew Skinner
Waste – Stephen Lee

Water – David Brewin

Upper Severn area
Tel: (01743) 272828
Fax: (01743) 272138
Environmental Planning Manager –
Robert Harvey
Environmental Protection Managers
– Carl Moss, David Sheldon

Lower Severn area
Tel: (01684) 850951
Fax: (01684) 293599
Environmental Planning Manager –
Roger Wade
Environmental Protection Manager –
Paul Quinn

Upper Trent area
Tel: (01543) 444141
Fax: (01543) 444161
Environmental Planning Manager –
David Hudson
Environmental Protection Manager –
Stuart Baker

Lower Trent area
Tel: (0115) 945 5722
Fax: (0115) 981 7743
Environmental Planning Manager –
Suzanne Davies
Environmental Protection Managers
– Jeff Dolby, Eric Stevens

Welsh region
HQ
Tel: (029) 2077 0088
Fax: (029) 2079 8555
IPC – Alun James
Pollution & Prevention Control –
Charlie Patterson
Waste – Martin Terry
Water – Kevin Thomas

Northern area
Tel: (01248) 670770
Fax: (01248) 670561
South-east area
Tel: (029) 2077 0088
Fax: (029) 2079 8555

South-west area
Tel: (01437) 760081
Fax: (01437) 760881

North-West region

HQ
Tel: (01925) 653999
Fax: (01925) 415961
IPC – Ian Haskell
Pollution & Prevention Control –
Roger Lamming
Water – Clive Gaskill

North area
Tel: (01228) 25151
Fax: (01228) 49734
Environmental Planning Manager –
Gerry McLoughlin
Environmental Protection Manager
– John Pinder

Central area
Tel: (01772) 39882
Fax: (01772) 627730
Environmental Planning Manager –
Alan D'Arcy
Environmental Protection Manager
– Laurence Rankin

South area
Tel: (0161) 973 2237
Fax: (0161) 973 4601
Environmental Planning Manager –
Stewart Lever
Environmental Protection Managers
– Roger Lamming, Dave Forster

National centres of expertise

Risk Analysis & Options Appraisal
Tel: (020) 7664 6811
Fax: (020) 7664 6911
Email:
gareth.llewellyn@environment-
agency.gov.uk
Gareth Llewellyn – Centre Head

**Ecotoxicology & Hazardous
Substances**
Tel: (01491) 832801
Fax: (01491) 834703
John Wade – Centre Head

Environmental Data & Surveillance
Tel: (01278) 457333
Fax: (01225) 469939
Dave Palmer – Centre Head

**National Groundwater &
Contaminated Land Centre**
Tel: (0121) 711 2324
Fax: (0121) 711 5925
Bob Harris – Centre Head

**National Water Demand
Management Centre**
Tel: (01903) 832000
Fax: (01903) 832274
Email: wdmc@dial.pipex.com
Peter Herbertson – Centre Head

National Centre for Coarse Fisheries
Tel: (01562) 863887
Fax: (01562) 69477
Email: phil.hinkley@environment-
agencygov.uk
Philip Hickley – Centre Head

**National Centre for Salmon &
Trout Fisheries Science**
Tel: (029) 2077 0088
Fax: (029) 2079 8383
Email: nigel.milner@environment-

agency.gov.uk
Nigel Milner – Centre Head
National Centre for Compliance Assessment
Tel: (01524) 842704
Fax: (01524) 842709
Email: stuart.newstead@environment-agency.gov.uk
Gareth Llewellyn – National Centre for Risk Analysis Options Appraisal

Scottish Environmental Protection Agency

Head office
Tel: (01786) 457700
Pollution reporting
Tel: (0345) 737271
Web: www.sepa.org.uk
Email: info@sepa.org.uk

North region
HQ
Tel: (01349) 862021
Fax: (01349) 863987
Aberdeen office
Tel: (01224) 248338
Fax: (01224) 248591

Elgin office
Tel: (01343) 547663
Fax: (01343) 540884

Fort William office
Tel: (01397) 704426
Fax: (01397) 705404

Fraserburgh office
Tel: (01346) 510502
Fax: (01346) 515444

Orkney office

Tel: (01856) 871080
Fax: (01856) 871090
Shetland office
Tel: (01595) 696926
Fax: (01595) 696946

Thurso office
Tel: (01847) 894422
Fax: (01847) 893365

Western Isles office
Tel: (01851) 706477
Fax: (01851) 703510

East region
HQ
Tel: (0131) 449 7296
Fax: (0131) 449 7277

Arbroath office
Tel: (01241) 874370
Fax: (01241) 430695

Galashiels office
Tel: (01896) 754797
Fax: (01896) 754412

Glenrothes office (air/waste)
Tel: (01592) 645565
Fax: (01592) 645567

Glenrothes office (water)
Tel: (01592) 759361
Fax: (01592) 759446

Perth office
Tel: (01738) 627989
Fax: (01738) 630997

Stirling office
Tel: (01786) 461407
Fax: (01786) 461425

West region

HQ
Tel: (01355) 238181
Fax: (01355) 264323

Ayr office
Tel: (01292) 264047
Fax: (01292) 611130

Dumfries office
Tel: (01387) 720502
Fax: (01387) 721154

Lochgilphead office
Tel: (01546) 602876
Fax: (01546) 602337

Newton Stewart office
Tel: (01671) 402618
Fax: (01671) 404121

Northern Ireland Environment and Heritage Service

Pollution reporting
Tel: (028) 9075 7414
Web: www.nics.gov.uk
Email: EHS@nics.gov.uk

Customer services
Tel: (028) 9054 6533

Information (leaflets/publications)
Tel: (028) 9054 6528

Conservation science
Tel: (028) 9054 6592

Conservation designations and protection
Tel: (028) 9054 6612

Countryside and coastal management
Tel: (028) 9054 6555

Regional operations
Tel: (028) 9054 6521

Industrial Air Pollution and Radiochemical Inspectorate
Tel: (028) 9025 4709

Drinking Water Inspectorate
Tel: (028) 9025 4862

Water Quality Unit
Tel: (028) 9025 4757

Environmental Quality Unit
Tel: (028) 9025 4816

Waste Management Inspectorate
Tel: (028) 9025 4815

Other Government and regulatory contacts

Department of the Environment, Transport & the Regions
Eland House
Bressenden Place
London SW1E 5DU
Tel: (020) 7890 3333
Publications orderline: (020) 8691 9191
Web: www.open.gov.uk

Department of Trade & Industry
1 Victoria Street
London SW1H 0ET
Tel: (020) 7215 5000
Publications orderline: (020) 7510 0174
Web: www.open.gov.uk

Energy Efficiency Office
See also Government Offices for the Regions for regional contacts in England

Environmental and Energy Management Directorate
Ashdown House
123 Victoria Street
London SW1E 6DE
Tel: (020) 7890 6655
Fax: (020) 7890 6689

Scotland
Scottish Office
Victoria Quay
Edinburgh EH6 6QQ
Tel: (0131) 244 7130
Fax: (0131) 244 7145

Wales
Cathays Park
Cardiff CF1 1NQ
Tel: (029) 2082 3126
Fax: (029) 2082 3661

Commission of the European Communities
England
Jean Monnet House
8 Storey's Gate
London SW1P 3AT
Tel: (020) 7973 1992

Scotland
9 Alva Street
Edinburgh EH2 4PH
Tel: (0131) 225 2058

Wales
4 Cathedral Road
Cardiff CF1 9SG
Tel: (029) 2037 1631

Northern Ireland
Windsor House
9/15 Bedford Street
Belfast BT2 7EG
Tel: (028) 9024 0708

European Environment Agency
Kongens Nytorv 6
DK-1050 Copenhagen K
Denmark
Tel: +45 33 36 71 00
Fax: +45 33 36 71 99
Email: eea@eea.dk

Web: www.eea.dk
UK national focal pont at the DETR
Tel: (020) 7276 8947

Drinking Water Inspectorate
Romney House
43 Marsham Street
London SW1P 3PY
Tel: (020) 7276 8296

Government Offices for the Regions
East Midlands
Belgrave Centre
Stanley Place
Talbot Street
Nottingham NG1 5GG
Tel: (0115) 971 9971
Fax: (0115) 971 2404

Eastern
Unit 7 Enterprise House
Vision Park
Chivers Way
Cambridge CB4 4ZR
Tel: (01234) 796332
Fax: (01234) 796252

London
Riverwalk House
157-161 Millbank
London SW1P 4RR
Tel: (020) 7217 3098
Fax: (020) 7217 3465

Merseyside
Cunard Building
Pier Head
Liverpool L1 1QB
Tel: (0151) 224 6300
Fax: (0151) 224 6471

North-East
Stanegate House
2 Groat Market

Newcastle upon Tyne NE1 1YN
Tel: (0191) 201 3300
Fax: (0191) 202 3806

North-West
2010 Sunley Tower
Piccadilly Plaza
Manchester M1 4BE
Tel: (0161) 952 4000
Fax: (0161) 952 4004

South-East
Bridge House
1 Walnut Tree Close
Guildford GU1 4GA
Tel: (01483) 882255
Fax: (01483) 882269

South-West
Fourth Floor
The Pithay
Bristol BS1 2PB
Tel: (0117) 900 1700
Fax: (0117) 900 1901

West Midlands
77 Paradise Circus
Queensway
Birmingham B1 2DT
Tel: (0121) 212 5000
Fax: (0121) 212 1010

Yorkshire & Humberside
Seventh Floor
East Wing
City House
Leeds LS1 4US
Tel: (0113) 280 0600
Fax: (0113) 244 9313

HSE Information Centre
Broad Lane
Sheffield S3 7HQ
Infoline: 0541 545500

HSE Books Tel: (01787) 881165
Fax: (01787) 313995
Web: www.open.gov.uk

Ministry of Agriculture, Fisheries & Food
Whitehall Place
London SW1A 2HH
Tel: 0645 335577
Web: www.open.gov.uk

Office of Electricity Regulation (OFFER)
Hagley House
Hagley Road
Edgbaston
Birmingham B16 8QG
Tel: (0121) 456 2100
Fax: (0121) 456 4664
Web: www.open.gov.uk

Office of Gas Supply (OFGAS)
Stockley House
130 Wilton Road
London SW1V 1LQ
Tel: (020) 7828 0898
Fax: (020) 7630 8164
Web: www.open.gov.uk

Office of Water Services (OFWAT)
Centre City Tower
7 Hill Street
Birmingham B5 4UA
Tel: (0121) 625 1300
Fax: (0121) 625 1400
Minicom: (0121) 625 1422
Web: www.open.gov.uk

Scottish Office
Pentland House
Robbs Loan
Edinburgh EH14 1TY
Tel: (0131) 556 8400
Web: www.open.gov.uk

Appendix 1

Pre-acquisition Due Diligence Inquiry

This Appendix is based on a precedent form produced by the Law Offices of NJF Lightbody, also available from www.aunet.co.uk/contaminated.

The proposed purchaser requests full and complete replies to these pre-contract enquiries from the vendor. Please provide a complete copy of any document referred to or, if a complete copy is not available, advise of the circumstances applicable.

PLEASE ENDORSE YOUR REPLIES ON THESE ENQUIRIES ON THE SAME PAGE AS THE ENQUIRY TO WHICH THE REPLY RELATES, SO FAR AS POSSIBLE.

The Law Offices of Nicholas J. F. Lightbody: proposed purchaser's solicitors

dated: _____

1. Environmental liability

With Statutory Guidance on liability for contaminated land due to come into force in mid 2000 it is necessary to request replies to enquiries designed to show whether or not there is cause for concern regarding the possibility of contamination of the property. We look forward to receiving replies setting out the vendor's knowledge of the property. Please see the attached definitions page (Edition 9.1 England & Wales) regarding the terms used in these environmental enquiries.

1.1 What was the land used for and what is it used for now?

(a) Current and proposed use

(i) Is the property or any Neighbouring property currently used for any Potentially contaminative use?

(ii) Is the seller aware of any proposal to use any Neighbouring property for any Potentially contaminative use?

(iii) Are any chemicals stored or located on the property excluding Retail Quantities purchased for private domestic use?

(b) Past use

(i) Has any environmental data report relating to the property from a recognised environmental data compiler been obtained?

(ii) Has the property or any Neighbouring property been used for any Potentially contaminative use?

(iii) Has the property been used

(A) for the storage of pesticides, automotive or industrial batteries, paints, or other chemicals (other than undamaged containers of consumer products of under five gallons in total volume)?

(B) Has the seller of the property been informed of the possible or actual existence of Hazardous substances on the property in the past?

(iv) Has there been a change in the nature of the use of, or activities carried on at, the property during the seller's ownership or occupation?

(v) Has the property ever been derelict land?

(c) General environmental liability

(i) Please confirm that no Hazardous substances have been used in the structure or fabric of any building on the property. If such confirmation cannot be given please give full details of the materials used and of any action taken to identify and remove any such materials from the property.

(ii) Has the seller of the property any reason for believing, suspecting or having some foundation for the belief that some potential liability or detriment arising from pollution or related environmental matters, whether of the property or Neighbouring property, may attach to the owners or occupiers of the property at any foreseeable future date?

(iii) Have any buildings on the property suffered from "sick building" syndrome or other related problems?

(iv) Have any buildings on the property been built in accordance with the Building Research Establishment Environmental Assessment Method (BREEAM)? If so please supply a copy of the assessment certificate.

1.2 Where is the land?

(a) Physical features

(i) Are there currently, or have there been previously, any Pits, Ponds or Lagoons located on the property in connection with waste treatment or waste disposal?

(ii) Is there currently, or has there been previously, any stained soil or significant evidence of damage to vegetation on the property?

(iii) Are there currently, or have there been previously, any storage tanks above or below ground located on the property?

(iv) Are there currently, or have there been previously, any vent pipes, fill pipes or access ways indicating a fill pipe protruding from the ground on the property or adjacent to any structure located on the property?

(v) Are there currently, or have there been previously, any flooring, drains, or walls located within the property that are stained by substances other than water or are emanating noxious odours?

(vi) Are there any "aggressive" ground conditions extant on the property where special protective measures are required to protect concrete or other underground structures and/or is there any extant chemical damage to underground telephone cables, electrical cables, fibre optic cables or other utility media?

(vii) Is/are there any electricity sub-station transformer capacitor or other oil-filled electrical switchgear located on the property or Neighbouring property for which there are no records indicating the absence of PCBs at significant concentrations?

(b) Investigations, Reports and Audits

(i) Have there been any Environmental audit soil or site investigations or samplings of the soil/water or atmosphere carried out on the property, whether by the local authority or any other person?

(ii) Has an Environmental audit, soil or site investigations or samplings, assessment or entry upon any statutory register relating to the property been carried out or made that indicates the presence or possible presence of Hazardous substances on the property, recommends further assessment of the property or indicates that the property had previously been used for any potentially contaminative use?

(iii) Has any environmental assessment (for planning, integrated pollution control or other purposes) been carried out in relation to the property or any proposed use of or development of the property?

(c) Water

(i) Is water abstracted for the benefit of the property? If so please supply copies of any abstraction licences and confirm that all conditions have been, and are being, complied with and that the seller is not aware of any proposals to vary or revoke any such licence or of any circumstances likely to lead to the variation or revocation of any such licence.

(ii) If water is abstracted for the benefit of the property, has the well/borehole or system been designated as contaminated by any statutory body?

(iii) If water is abstracted for the benefit of the property has the abstracted water been subject to chemical analysis?

(iv) Has the ground water or any aquifer beneath the property or any Neighbouring property been contaminated by Hazardous substances, sewage or any other substance being a known or potential hazard to health?

(v) Does the property discharge waste water, other than storm water, directly to a ditch or stream on or adjacent to the property or to a private or public sewer?

(vi) Where storm waters include surface drainage from areas of hardstanding is there an appropriate degree of protection of discharge water purity provided by oil interceptors?

(vii) Please confirm that no polluting incident (including, but not limited to, the entry of Hazardous substances into any surface or ground water or public or private sewer) has taken place, or is taking place on the property. If such an incident has taken, or is taking, place please provide the following particulars:

(A) full details of the accident or incident;

(B) copies of any reports, correspondence, court orders, notices or recommendations relating to the accident or incident;

(C) details of any remedial work carried out, including certificates of satisfactory completion.

(viii) Is the seller aware of any matter which may adversely affect the potable quality of the water supply to the property (such as, inter alia, lead pipes and the presence of pathogens).

(d) Potentially contaminated material or fill

(i) Has hard-core, rubble, soil, sand or any other material been brought onto the property which originated from a contaminated site or is of an unknown origin?

(ii) Have any Construction debris, Hazardous substances, unidentified waste materials, tyres, automotive or industrial batteries or any other waste materials, rubbish, debris or refuse been dumped above ground, buried and/or burned on the property?

(iii) Please confirm that the land upon which any buildings and structures on the property have been constructed is not reclaimed, "made" or filled land.

(iv) Have any current or past buildings on the property included Asbestos containing material (inter alia as roofing, cladding or insulation)?

(v) Has any urea formaldehyde foam insulation been used within the fabric of any building on the property?

1.3 Does any statutory or non-statutory register or database contain any relevant entry?

(a) Actual or threatened litigation or proceedings

(i) Is there any past, threatened or pending litigation or administrative proceedings concerning any release or threatened release of any Hazardous substance involving the property whether by the seller, by any other party with a legal interest in the property or by any other party with a legal interest in any Neighbouring property?

(ii) Is the seller aware of any actual, intended or possible proceedings by an aggrieved person under s.82 of the Environmental Protection Act 1990 or any equivalent legislation in relation to any matters affecting the property or any Neighbouring property?

(b) Statutory matters: consents

(i) Please provide a schedule of all existing consents permissions and authorisation ("consent") from any statutory body for atmospheric emissions, trade effluent discharges and disposal of solid wastes to tips or landfills on the property together with copies of all relevant documents.

(NOTE: Ensure contract provides that such consents will be transferred to the purchaser and check that any conditions to which such consents are subject will not become more stringent upon the regulatory authorities recognising the new owners of the property).

(ii) If any processes carried on at the property are prescribed processes within the meaning of the Prescribed Processes and Substances Regulations (as amended), or involve emissions to any environmental medium from the property which are prescribed substances within the meaning of those Regulations, please supply a copy of any consent, including a copy of any application of such authorisation or consent where the authorisation or consent is pending or has been refused.

(iii) Please confirm that all conditions attached to such consents have been and are being complied with, and that the seller is not aware of any proposals to vary or revoke any such consent or of any circumstances likely to lead to the variation or revocation of any such consent.

(iv) Please give details of any appeals or other proceedings, discussions or negotiations with any relevant body with respect to the grant, revocation, renewal or variation of any contract, agreement, authorisation, consent, licence or arrangement relating to matters of environmental concern.

(c) Statutory matters: breach

(i) Does the seller of the property have any knowledge of any statutory notification relating to past or current breach of health and safety or environmental statutes or subordinate legislation with respect to the property?

(ii) Has the seller been advised of the breach of any health and safety or environmental statute or subordinate legislation with respect to the property?

(iii) Have any notices been served in respect of the property under either s.58 or s.60 of the Control of Pollution Act 1974? If so, please supply copies of any such notices and full details of the result of any subsequent proceedings.

(iv) Is the seller aware of any circumstances or proposals to serve any notice or make or declare any order under the Control of Pollution Act 1974, the Health and Safety at Work Act 1974 or the Environmental Protection Act 1990?

(v) Is there any reason for the Environment Agency to carry out or is the seller aware of any work having been carried out or intended to be carried out by the Environment Agency pursuant to s.161 of the Water Resources Act 1991?

(vi) Is the seller aware of or have there been any complaints from or disputes with the local authority, any other statutory authority or any other person regarding the state and condition of the property insofar as it affects the environment by reason of the emission of any substance into the air, water or land from the property or in relation to noise, odours, heat, light or radiation? In particular, have any remediation, works or charging notices been served?

(vii) Is there any reason why the relevant waste regulation authority might enter or inspect the property or take any steps to avoid pollution to the environment or harm to health, or any indication that the authority might do any of those things?

(d) Identification of contaminated land & special sites under s.78 EPA 1990

(i) Has the Environment Agency/Local Authority inspected the land or given notice of its intended inspection of the land?

(ii) Is the seller of the property aware of any intention or possible intention on the part of the Environmental Agency/Local Authority to identify the property or any Neighbouring property as contaminated land or a "special site" within the meaning of Part IIA Environmental Protection Act 1990 (as amended by the Environment Act 1995) or any similar provision?

(e) Pollution control on industrial premises

If the property includes any operative industrial facilities what is the size of existing capital, operating and maintenance budgets for existing new or improved pollution control equipment on the property?

(f) Waste Management

(i) How is all waste on the property (other than normal domestic waste) or other materials, including trade effluent, stored or disposed of?

(ii) Is the property or any Neighbouring property the subject of a waste management licence or of any regulations made by the Secretary of State for the Environment exempting such property from a requirement to have such a licence?

(iii) Please supply copies of all necessary permissions, licences, contracts, or arrangements relating to the disposal of trade effluent, commercial waste or industrial water or relating to the removal and disposal of substances produced in the course of treating any trade effluent, commercial waste or industrial waste.

(iv) Please confirm that the obligations contained in any licence, contract or arrangement relating to the disposal of trade effluent commercial waste or industrial waste or relating to the removal and disposal of substances produced in the course of treating any trade effluent commercial waste or industrial waste are being, and have at all times been, complied with and that the seller is not aware of any proposals or circumstances which might give rise to the termination of any such licence, contract or arrangement.

(g) Air Pollution

Has compliance been made at all times, and in all respects with the requirements of the Clean Air Act 1993?

(h) Radioactive substances

(i) Has there been any use, storage or disposal of radioactive substances on the property?

(ii) Is radon gas known to be emitted from beneath the property?

(iii) Are there any radioactive lightning conductors on the property?

(iv) Has the seller knowledge of any investigations having been carried out on the property in relation to the possible presence of radio-active substances and/or radon gas?

(i) Noise

Is the property within a noise abatement zone?

(j) Electromagnetic Radiation

Does the property lie within 150 metres of any overhead or underground electricity cables transmitting electrical power in excess of 132,000 volts?

(k) Insurance

(i) Does the seller of the property have the benefit of any insurance policy or policies providing cover against Third Party and/or Public Liability risks arising on the property and in particular do any "occurrence" based policies exist, of whatever date?

(ii) Does the property have the benefit of any environmental insurance?

(iii) Has an application ever been made for such environmental insurance cover, whether or not it was successful?

(iv) Does the property have the benefit of any guarantee warranty or insurance policy in respect of environmental liability, loss of value or defects resulting from past contaminative use.

Definitions and Interpretations for use with ENVIRONMENTAL PRE-CONTRACT AND DUE DILIGENCE ENQUIRIES edition 9.1 (England & Wales)

1. Definitions

1.1 "Asbestos" – naturally occurring fibrous materials found in certain types of rock formations including, inter alia, the minerals chrysotile, amosite and crocidolite.

1.2 "Asbestos containing material" – any material or product which contains more than 1% Asbestos.

1.3 "Construction debris" – concrete, brick, asphalt and other such inert materials discarded in the construction or demolition of an improvement to property.

1.4 "Environmental audit" – an investigative process to determine, inter alia, if an existing facility or company is in compliance with the applicable environmental laws and regulations or an approved corporate environmental policy.

1.5 "Hazardous substance" or "Hazardous substances" –

(a) any substance or substances referred to in EC Directive 76/464/EEC List I and List II of families and groups of substances as follows:

(i) List I contains certain individual substances which belong to the following families and groups of substances, selected mainly on the basis of their toxicity, persistence and bioaccumulation, with the exception of those which are biologically harmless or which are rapidly converted into substances which are biologically harmless:

(A) organohalogen compounds and substances which may form such compounds in the aquatic environment,

(B) organophosphorus compounds,

(C) organotin compounds,

(D) substances in respect of which it has been proved that they possess carcinogenic properties in or via the aquatic environment,

(E) mercury and its compounds,

(F) cadmium and its compounds,

(G) persistent mineral oils and hydrocarbons of petroleum origin, and

(H) persistent synthetic substances which may float, remain in suspension or sink and which may interfere with any use of water

(ii) List II contains certain individual substances and categories of substances belonging to the following families and groups of substances which have a deleterious effect on the aquatic environment:

(A) The following metalloids and metals and their compounds:

zinc copper   nickel   chromium
lead selenium arsenic   antimony
molybdenum titanum tin   barium
beryllium   boron   uranium vanadium
cobalt   thallium tellurium silver

(B) Biocides and their derivatives not appearing in List I.

(C) Substances which have a deleterious effect on the taste and/or smell of the products for human consumption derived from the aquatic environment, and compounds liable to give rise to such substances in water.

(D) Toxic or persistent organic compounds of silicon, and substances which may give rise to such compounds in water, excluding those which are biologically harmless or are rapidly concerted in water into harmless substances.

(E) Inorganic compounds of phosphorus and elemental phosphorus.

(F) Non-persistent mineral oils and hydrocarbons of petroleum origin.

(G) Cyanides, fluorides.

(H) Substances which have an adverse effect on the oxygen balance, particularly: ammonia, nitrates.

(b) and/or substances controlled by Directives created pursuant to the said Directive 76/464/EEC together with Asbestos and any other substance compound or element generally regarded as being toxic and/or harmful and/or hazardous to health and/or

(c) "Noxious gases" – those gases which by virtue of their physical or chemical properties are considered to be toxic, carcinogenic, irritant, asphyxiant, flammable or explosive and may pose a risk to human health or the built environment, including inter alia: methane (and landfill gas), carbon monoxide, carbon dioxide, hydrogen cyanide, hydrogen sulphide, phosphine, sulphur dioxide, formaldehyde and radon.

1.6 "Licensed" or "License" written consent from a statutory body.

1.7 "Neighbouring property" – any land lying within 500 metres from the location by which the property has been defined.

1.8 "Potentially contaminative use" – past uses of land which may fall within the following categories:

(a) Agriculture: burial of diseased livestock.

(b) Extractive industry: coal mines, coal preparation plants, oil refineries and petrochemicals; mineral workings, mineral processing works. (NB: includes loading, transportation, sorting, forming and packaging and similar operations.)

(c) Energy industry: gas works; coal carbonisation plants; oil refineries; power stations.

(d) Production of metals: metal processing; heavy engineering; electroplating and metal finishing. (NB: includes scrap metal treatments.)

(e) Production of non-metals and their products: mineral processing; asbestos works; cement, lime and gypsum manufacture, brickworks and associated processes.

(f) Glass-making and ceramics – including glazes and vitreous enamel.

(g) Production and use of chemicals.

(h) Engineering and manufacturing processes: manufacture of metal goods (including mechanical engineering industrial plant or steel-work, motor vehicles, ships, railway or tramway vehicles, etc.); storage, manufacture or testing of explosives, propellants, small arms, etc.; electrical and electronic equipment manufacture and repair.

(i) Food processing industry: pet foods or animal feedstuffs; processing of animal by-products (including rendering or maggot farming but excluding slaughterhouses and butchering).

(j) Paper, pulp and printing industry.

(k) Timber and timber products industry: chemical treatment and coating of timber and timber products.

(l) Textile industry: tanning, dressing fellmongering or other process for preparing, treating or working leather; fulling, bleaching, dyeing or finishing fabrics or fibres; manufacture of carpets or other textile floor-coverings (including linoleum works).

(m) Rubber industry: processing natural or synthetic rubber (including tyre manufacture or retreading).

(n) Infrastructure: marshalling, dismantling, repairing or maintenance of railway rolling-stock; dismantling, repairing or maintenance of marine vessels (including hovercraft); dismantling, repairing or maintenance of road transport or road haulage vehicles; dismantling, repairing or maintenance of air and space transport systems.

(o) Waste disposal: treating sewage or other effluent or storage, treatment or disposal of sludge (including sludge from water treatment work); treating, keeping, depositing or disposing of waste including scrap (to include landfill, infilled canal basins, docks or river courses); storage or disposal of radioactive materials.

(p) Miscellaneous: dry-cleaning operations; laboratories for educational or research purposes; demolition of buildings, plant or equipment for any of the activities mentioned above.

1.9 "Prescribed Process" processes defined in the Environmental Protection (Prescribed Processes and Substances) Regulations 1991 (as amended).

1.10 "Reliable Publicly Accessible Mapping" means all of the following: a map from the Ordnance Survey County Series published pre-1900, a map from the same series published between 1900 and 1919 (inclusive), a map from the same series published between 1920 and 1945 (inclusive) and a map published by Ordnance Survey in their National Grid Series post-1945.

1.11 "Retail Quantities" materials or substances purchased from normal retail sources in any normal retail quantity excluding fuel for internal combustion engines and any heating system.

2. Interpretation

2.1 In these enquiries the purchaser requires the seller to reveal the seller's knowledge of the property and of Neighbouring property. The seller is assumed by the purchaser to possess the level of knowledge of the property, and of Neighbouring property, that is reasonably to be expected of a prudent and well advised purchaser who has already undertaken pre-contract enquiries and the relevant searches and investigations regarding the property and Neighbouring property.

2.2 These enquiries refer throughout to "seller" and "purchaser" being the parties respectively disposing of and acquiring their interest in the property. These terms shall be deemed to include reference to "mortgagor" and "mortgagee" and to "landlord" and "tenant" respectively as may be relevant in the circumstances without modification.

2.3 Where any substantive matter is revealed by these enquiries the seller is requested to provide full details to the purchaser.

2.4 The singular shall be deemed to include the plural, the plural to the include the singular and the male to include the female and neuter.

IPC and Air Pollution Guidance

All guidance notes are available from Stationery Office Books (Tel: (020) 7873 0011)

This Appendix details the present Chief Inspectors/Integrated Pollution Guidance notes.

Note that the Environment Agency is in the process of revising all guidance published by the former HMIP. This second series is known as IPC guidance and is distinguishable by the prefix S2.

Fuel production processes

S2 1.01	Combustion processes: boilers & furnaces >50MW supersedes IPR 1/1
IPR1/2	Combustion processes: gas turbines
S2 1.03	Combustion processes: compression ignition engines >50MW supersedes IPR 1/3
S2 1.04	Combustion processes: waste & recovered oil burners >3MW supersedes IPR 1/4
S2 1.05	Combustion processes: solid fuel made from municipal waste >3MW supersedes IPR 1/5-1/8
S2 1.06	Carbonisation & associated processes: coke manufacture supersedes IPR1/9
S2 1.07	Carbonisation & associated processes: smokeless fuel, activated carbon & carbon black manufacture supersedes IPR 1/10
S2 1.08	Gasification processes: gasification of solid & liquid feedstocks supersedes IPR 1/11
S2 1.09	Gasification processes: refining of natural gas supersedes IPR 1/12-1/3
S2 1.10	Gasification processes: oil refining and associated processes supersedes IPR 1/14-1/15
S2 1.11	Petroleum processes: on-shore oil production supersedes IPR 1/16

S2 1.12   Combustion processes: reheat & heat treatment furnaces supersedes IPR 1/17

Metals production and processing

IPR2/1	Iron & steel making processes: integrated iron & steel works
IPR2/2	Ferrous foundry processes
IPR2/3	Processes for electric arc steelmaking, secondary steelmaking & special alloy production
IPR2/4	Processes for the production of zinc & zinc alloys
IPR2/5	Processes for the production of lead & lead alloys
IPR2/6	Processes for the production of refractory metals
IPR2/7	Processes for the production, melting & recovery of cadmium, mercury & their alloys
IPR2/8	Processes for the production of aluminium
IPR2/9	Processes for the production of copper & copper alloys
IPR2/10	Processes for the production of precious metals & platinum group metals
IPR2/11	The extraction of nickel by the carbonyl process & the production of cobalt & nickel alloys
IPR2/12	Tin & bismuth processes

Mineral industry sector

S2 3.01	Cement manufacture , lime manufacture & associated processes supersedes IPR 3/1-3/2
S2 3.02	Asbestos processes supersedes IPR 3/3
S2 3.03	Manufacture of glass fibres, other non-asbestos mineral fibres, glass frit, enamel frit & associated processes supersedes IPR 3/4-3/5
S2 3.04	Ceramic processes supersedes IPR 3/6

Chemical industry sector

IPR4/6	Production & polymerisation of organic monomer
IPR4/14	Processes for the manufacture, use or release of hydrogen halides or any of their acids
IPR4/16	The manufacture of chemical fertilisers or their conversion into granules
IPR4/17	Bulk storage installations
IPR4/18	Processes for the manufacture of ammonia

IPR4/20	The production & use of, in any chemical manufacturing process, phosphorus & any oxide, hydride, or halide of phosphorus
IPR4/21	Processes involving the manufacture, use or release of hydrogen cyanide or hydrogen sulphide
IPR4/22	Processes involving the use or release of antimony, arsenic, beryllium, gallium, indium, lead, palladium, platinum, selenium, tellurium, thallium or their compounds
IPR4/23	Processes involving the use or release of cadmium or any cadmium compound
IPR4/24	Processes involving the use or release of mercury or any mercury compound
IPR4/25	Processes for the production of compounds of chromium, magnesium, manganese, nickel & zinc

Waste disposal and recycling sector

S2 5.01	Waste incineration supersedes IPR 5/1-5/5, 5/11
S2 5.02	Making solid fuel from waste supersedes IPR 5/6
S2 5.03	Cleaning & regeneration of carbon supersedes IPR 5/7
S2 5.04	Recovery of organic solvents & oil by distillation supersedes IPR 5/8, 5/10
IPR5/9	Regeneration of ion exchange resins

Other industries

IPR6/1	Application or removal of tributyltin or triphenyltin at ship/boatyards
IPR6/2	Tar & bitumen processes
IPR6/3	Timber preservation processes
IPR6/4	Di-isocyanate manufacture
IPR6/5	Toluene di-isocyanate use & flame bonding of polyurethanes
IPR6/6	Textile treatment processes
IPR6/7	Processing of animal hides & skin
IPR6/8	The making of paper pulp by chemical methods
IPR6/9	Paper making & related processes, inc mechanical pulping, recycled fibres & de-inking

Process guidance notes

Process guidance noted relating specifically to LAAPC processes are issued by the DETR. These are detailed as follows:

Fuel production processes

PG1/1(95)	Waste oil burners <0.4MW
PG1/2(95)	Waste oil burners <3MW
PG1/3(95)	Boilers & furnaces 20-50MW
PG1/4(95)	Gas turbines 20-50MW
PG1/5(95)	Compression ignition engines 20-50MW
PG1/6(91)	Tyre & rubber combustion 0.4-3MW superseded by PG1/12(95)
PG1/7(91)	Straw combustion 0.4-3MW superseded by PG1/12(95)
PG1/8(91)	Wood combustion 0.4-3MW superseded by PG1/12(95)
PG1/9(91)	Poultry litter combustion 0.4-3MW superseded by PG1/12(95)
PG1/10(92)	Waste derived fuel combustion <3MW
PG1/11(96)	Reheat & heat treatment furnaces 20-50MW
PG1/12(95)	Fuel made from solid waste combustion 0.4-3MW
PG1/13(96)	Processes for the storage, loading and unloading of petrol at terminals
PG1/14(96)	Unloading of petrol into storage at service stations

Metals production and processing

PG2/1(96)	Furnaces for extracting non-ferrous metal from scrap
PG2/2(96)	Hot dip galvanising
PG2/3(96)	Electrical & rotary furnaces
PG2/4(96)	Iron, steel & non-ferrous metal foundries
PG2/5(96)	Hot & cold blast cupolas
PG2/6(96)	Aluminium & aluminium alloy processes
PG2/7(96)	Zinc & zinc alloy processes
PG2/8(96)	Copper & copper alloy processes
PG2/9(96)	Metal decontamination processes

Minerals industry

PG3/1(95)	Blending, packing, loading, use of bulk cement
PG3/2(95)	Manufacture of heavy clay goods & refractory goods
PG3/3(95)	Glass (not lead glass) manufacturing processes

PG3/4(95)	Lead glass manufacturing processes
PG3/5(95)	Coal, coke, coal product & petroleum coke processes
PG3/6(95)	Polishing or etching of glass using hydrofluoric acid
PG3/7(95)	Vermiculite exfoliation & expansion of perlite
PG3/8(96)	Quarry processes
PG3/9(91)	Sand drying & cooling superseded by PG3/15(96)
PG3/10(91)	China & ball clay superseded by PG3/17(95)
PG3/11(91)	Spray drying of ceramic materials superseded by PG3/17(95)
PG3/12(95)	Plaster processes
PG3/13(95)	Asbestos processes
PG3/14(95)	Lime processes
PG3/15(96)	Mineral drying & roadstone coating
PG3/16/96	Mobile screening & crushing
PG3/17(95)	China & ball clay processes inc spray drying

Chemicals industry

| PG4/1(94) | Surface treatment of metals |
| PG4/2(96) | Fibre reinforced plastics manufacture |

Waste disposal and recycling

PG5/1(95)	Clinical waste incineration <1t/hr
PG5/2(95)	Crematoria
PG5/3(95)	Animal carcase incineration processes <1t/hr
PG5/4(95)	General waste incineration processes <1t/hr
PG5/5(91)	Sewage incineration

Other industries

PG6/1(91)	Animal by product rendering
PG6/2(95)	Timber & wood-based products manufacture
PG6/3(97)	Chemical treatment of timber & wood-based products
PG6/4(95)	Particleboard & fibreboard manufacture
PG6/5(95)	Maggot breeding
PG6/7(97)	Printing & coating of metal packaging
PG6/8(97)	Textile & fabric coating & finishing
PG6/9(96)	Coating powder manufacture
PG6/10(97)	Coating manufacturing
PG6/11(97)	Printing ink manufacture
PG6/12(91)	Production of natural sausage casings, tripe, chitterlings etc

PG6/13(97)	Coil coating
PG6/14(97)	Film coating
PG6/15(97)	Coating in drum manufacturing & reconditioning
PG6/16(97)	Printworks
PG6/17(97)	Printing of flexible packaging
PG6/18(97)	Paper coating
PG6/19(97)	Fish meal & fish oil
PG6/20(97)	Paint application in vehicle manufacturing
PG6/21(96)	Hide & skin processes
PG6/22(97)	Leather finishing
PG6/23(97)	Metal & plastic coating
PG6/24(96)	Pet food manufacture
PG6/25(92)	Vegetable oil extraction & fat refining
PG6/26(96)	Animal feed compounding
PG6/27(96)	Vegetable matter drying
PG6/28(97)	Rubber processes
PG6/29(97)	Di-isocyanate processes
PG6/30(97)	Mushroom compost production
PG6/31(96)	Powder coating (inc sheradizing)
PG6/32(97)	Adhesive coating
PG6/33(97)	Wood coating
PG6/34(97)	Road vehicles respraying
PG6/35(96)	Metal & other thermal spraying
PG6/36(97)	Tobacco processing
PG6/38(92)	Blood processing
PG6/39(92)	Animal by-product dealers
PG6/40(94)	Coating & recoating of aircraft & components
PG6/41(94)	Coating & recoating of rail vehicles
PG6/42(94)	Bitumen & tar processes

Other IPC and air pollution guidance

General guidance notes

The former DoE issued general guidance on the main EPA 1990 controls and procedures.

GG1	Introduction to EPA 1990 part I (explanation of BATNEEC, "substantial" changes, existing process, the charging scheme, variation notices)
GG2	Authorisations advice for LAs
GG3	Applications and registers (includes contact details of the statutory consultees)

GG4	Interpretation of terms used in Pollution Guidance Notes
GG5	Appeals

Upgrading guidance

UG1	Revisions/additions to existing Pollution Guidance and General Guidance Notes affects PGs 1/1, 1/3, 1/5, 1/10, 3/2, 5/1, 5/2, 5/4, 5/5, 6/3, 6/21, 6/22, 6/23, 6/26 and gives additional general guidance

Technical guidance notes

Monitoring

M1	Sampling facility requirements for monitoring particulates in gaseous releases
M2	Monitoring emissions of pollutants at source
M3	Standards for IPC monitoring part 1: standards organisations and measurement infrastructure
M4	Standards for IPC monitoring part 2: standards in support of IPC monitoring

Dispersion

D1	Guidelines on discharge stack heights for polluting emissions

Abatement

A1	Flaring in the gas, petroleum, petrochemical & associated industries
A2	Pollution abatement technology for the reduction of solvent vapour emissions
A3	Pollution abatement technology for particulate and trace gas removal
A4	Effluent treatment techniques

Environment

E1	Best practicable environmental option assessments for Integrated Pollution Control

· **Proposed Phase-in Dates** ·

The following table shows the proposed phase-in dates for various industrial catergories under IPPC with BREF production start date.

Sector and BRIEF Note production	Phase-in date for Part A (1) installation (E&W)	Phase-in date for Part A (2) installation (E&W)	Phase-in date for Part A installation (SCO)
Paper/pulp (1997)	2000	2001	2001
Primary/secondary steel (1997)			
Textiles (1998)			
Tanneries (1998)			
Cement and lime (1997)			
Ferrous metal processing (1998)	2001	2002	2002
Non-ferrous metal production and processing (1998)			
Glass (1998)			
Chloralkali 1998			
Smitheries and foundries (1999)	2002	2003	2003
Large volume organic without batch processes (1999)			
Food and milk (2000)			
Livestock and poultry (1999)	2003	2004	2003
Asbestos (2001)			
Ceramics (2001)			
Polymers (2001)			
Large volume solid organic (2000)			
Slaughterhouses/carcasses (2000)			
Surface treatment of metals (2001)			
Landfills (begin phasing in) (2002)			
Pigs (1999)	2004	2005	2004
Hazardous waste incineration (2000)			
Municipal waste incineration (2002)			
Waste disposal and recovery (2002) (other than landfill and incineration)			

Sector and BRIEF Note production	Phase-in date for Part A (1) installation (E&W)	Phase-in date for Part A (2) installation (E&W)	Phase-in date for Part A installation (SCO)
Batch organics in multi-purpose plant (1999)	2005	2006	2005
Large volume gas and liquid inorganic (1999)			
Speciality organics (2002)	2006	2007	2006
Organic fine chemicals (2002)			
Coating activities using organic solvents (2001)			
Refineries (1999)			
Large combustion plant (2001)			
Coal liquefaction (2001)	2007	2007	2007

Appendix 4

DoE List of Contaminative Uses

C.1 **Agriculture**

a.   Burial of diseased livestock

C.2 **Extractive Industry**

a   Extracting, handling and storage of carbonaceous materials such as coal, lignite, petroleum, natural gas, or bituminous shale (not including the underground workings).
Profiles: coal mines and coal preparation plants, oil refineries and petrochemicals.

b.   Extracting, handling and storage of ores and their constituents.
Profiles: mineral workings, mineral processing works.
Note: Handling includes loading, transport, sorting, forming and packaging, and similar operations. Ore means any mineral, including non-metal bearing, except fuels.

C.3 **Energy Industry**

a.   Producing gas from coal, lignite, oil or other carbonaceous material (other than from sewage or other waste), or from mixtures of those materials.
Profiles: Gasworks and coal carbonisation plants; oil refineries.

b.   Reforming, refining, purifying and odourising natural gas or any product of the processes outlined in C.3a above.
Profiles: Gasworks and coal carbonising plants; oil refineries.

c.   Pyrolysis, carbonisation, distillation, liquefaction, partial oxidation, other heat treatment, conversion, purification, or refining or coal, lignite, oil, other carbonaceous material of mixtures and products thereof, otherwise than with a view to gasification or making of charcoal.
Profiles: Gasworks and coal carbonisation plants; oil refineries; coal mines and coal preparation plants.

d. A thermal power station (including nuclear power stations and production, enrichment and reprocessing of nuclear fuels).
    Profiles: power stations; radioactive materials; asbestos works.

e. Electricity and sub-station.
    Profiles: power stations; electrical equipment.

C.4 Production of Metals

a. Production, refining or recovery of metals by physical, chemical, thermal or electrolytic or other extracting process.
    Profiles: Metal processing; heavy engineering.

b. Heating, melting or casting metals as part of an intermediate or final manufacturing process (including annealing, tempering or similar processes).
    Profiles: metal processing; heavy engineering; miscellaneous (High Street) trades.

c. Cold forming processes (including pressing, rolling, extruding, stamping, forming or similar processes).
    Profiles: metal processing; heavy engineering; electroplating and metal finishing.

d. Finishing treatments, including anodising, pickling, coating, and plating or similar processes.
    Profiles: metal processing; heavy engineering; electroplating and metal finishing; miscellaneous (High Street) trades.
    Note: metals are taken to include metal scrap.

C.5 Production of Non-Metals and their Products

a. Production or refining of non-metals by treatment of the ore.
    Profiles: mineral processing works.

b. Production or processing of mineral fibres by treatment of the ore.
    Profile: mineral processing works; asbestos works.

c. Cement, lime and gypsum manufacture, brickworks and associated processes.
    Profile: mineral processing works.

C.6 Glass Making and Ceramics

a. Manufacture of glass and products based on glass.

Profile: glass manufacturing.

b. Manufacture of ceramics and products based on ceramics, including glazes and vitreous enamel.

C.7 **Production and Use of Chemicals**

a. Production, refining, recovery or storage of petroleum or petrochemicals, or their by-products, including tar and bitumen processes and manufacture of asphalt.
Profiles: oil refineries and petrochemicals; mineral processing works; drum and tank cleaning.

b. Production, refining and bulk storage of organic or inorganic chemicals, including fertilisers, pesticides, pharmaceuticals, soaps, detergents, cosmetics, toiletries, dyestuffs, inks, paints, fireworks, pyrotechnic materials or recovered chemicals.
Profiles: bulk inorganic and organic chemicals; pesticides, pharmaceuticals; textile and dye industry; paint and ink manufacture; miscellaneous (High Street) trades.

c. Production, refining and bulk storage of industrial goods not otherwise covered.
Profile: fine chemicals.

C.8 **Engineering and Manufacturing Processes**

a. Manufacture of metal goods, including mechanical engineering, industrial plant or steel works, motor vehicles, ships, railway or tramway vehicles, aircraft, aerospace equipment or similar equipment.
Profiles: heavy engineering works; engineering works; car manufacturing works; shipbuilding.

b. Storage, manufacture or testing of explosives, propellants, ordnance, small arms or ammunition.
Profile: heavy engineering.

c. Manufacture and repair of electric and electronic components and equipment.
Profiles: electrical and electronic equipment manufacture; miscellaneous (High Street) trades.

C.9 **Food Processing Industry**

a. Manufacture of pet foods or animal foodstuffs.

Profile: food preparation and processing.

b. Processing of animal by-products (including rendering or maggot farming, but excluding slaughterhouses, butchering).
    Profiles: Animal processing works; miscellaneous (High Street) trades.

C.10 **Paper, Pulp and Printing Industry**

a. Making of paper pulp, paper or board, or paper or board products, including printing or de-inking.
    Profiles: pulp and paper manufacture; printing works; miscellaneous (High Street) trades.

C.11 **Timber and Timber Products Industry**

a. Chemical treatment and coating of timber and timber products.
    Profiles: wood preservative industry and timber treatment works; miscellaneous (High Street) trades.

C.12 **Textile Industry**

a. Tanning, dressing, fell mongering or other process for preparing, treating or working leather.
    Profiles: animal processing works; miscellaneous (High Street) trades.

b. Fulling, bleaching, dyeing or finishing fabrics or fibres.
    Profiles: textile and dye industry; miscellaneous (High Street) trades.

c. Manufacture of carpets or other textile floor coverings (including linoleum works).
    Profile: textile and dye industry.

C.13 **Rubber Industry**

a. Processing of natural or synthetic rubber (including tyre manufacture or retreading).
    Profiles: fine chemicals; tyre manufacture.

C.14 **Infrastructure**

a. Marshalling, dismantling, repairing or maintenance of railway rolling stock.

Profiles: heavy engineering; docks and railway land.

b. Dismantling, repairing or maintenance of marine vessels, including hovercraft.
Profiles: shipbuilding and ship breaking; docks and railway land.

c. Dismantling, repairing or maintenance of road transport and road haulage; garages and filling stations.
Profiles: road transport and road haulage; garages and filling stations.

d. Dismantling, repairing or maintenance of air or space transport systems.
Profiles: engineering works; airports.

C.15 **Waste Disposal**

a. Treating of sewage or other effluent.
Profile: sewage works and farms.

b. Storage, treatment or disposal of sludge including sludge from water treatment works.

c. Treating, keeping, depositing or disposing of waste, including scrap (to include infilled canal basins, docks or river courses).
Profiles: landfills and other waste treatment and disposal sites; scrap yards; drum and tank cleaning.

d. Storage or disposal of radioactive materials.
Profile: radioactive materials.

C.16 **Miscellaneous**

a. Premises housing dry cleaning operations.
Profile: miscellaneous (High Street) trades.

b. Laboratories for educational or research purposes.
Profiles: research laboratories; miscellaneous (High Street) trades.

c. Demolition of buildings; plant or equipment used for any of the activities in this schedule.
Profile: demolition.

· Pre-survey Questionnaire ·

As part of an overall auditing programme for [insert site name], Environmental Auditors Limited have been asked to undertake an Environmental Audit of your site. Prior to us visiting the site however it would be beneficial to us if you could answer the following questions.

Name of site:

Address of site:

Nature of company's business:

Principal processes operated at site:

Area of site:

Number of employees:

Operational hours:

Who at the facility is responsible for the following (please give contact numbers or extension numbers)?:

•   Overall Environmental Management:

•   Air Pollution Control:

- Waste Management:

- Drainage and Waste Water Disposal:

Does the site have an Environmental Policy?:

If Yes please attach a copy of the policy when returning this form.

Does the site hold any of the following consents, authorisations or licenses? For those which are held please give an indication of the nature for the consent, authorisation or licence:

- Part A Process Authorisation (IPC Authorisation)

- Part B Process Authorisation (APC Authorisation)

- Waste Management Licence

- Controlled Water Discharge Consent

- Trade Effluent Discharge Consent

- Waste Carrier Certification

- Hazardous Substances Consent

- Radioactive Substances Registration

Please send this form back to:
ENVIRONMENTAL AUDITORS LIMITED
EAL House, Red House Farm, Newtimber, Hassocks, West Sussex, BN6 9BS
Tel.: 01273 857500 Fax.: 01273 857550   e-Mail: info@environmentalauditors.co.uk
Web: environmentalauditors.co.uk

Document Request Form

As part of the audit process it will be necessary for us to inspect any relevant environmental documentation pertaining to your site and the operations undertaken upon it. As such, and in order to minimise the time spent on site, could you please arrange to have the following documentation listed below available for inspection during the audit. The following list is relatively comprehensive and it is unlikely that you will have all of it. If you have any queries about the list please contact us on (01273) 857500.

The documentation that we will need to see is as follows:

- A plan of the site;
- The company's environmental policy;
- Details and Documentation relating to any Environmental Management System that you may have in place;
- Evidence of any Environmental Training that may be given to staff (such as records as to who has received training, course materials etc.);
- Any environmental audit, review or inspection reports that you may have had undertaken previously;
- Any contaminated land surveys, assessments or investigations that may have been undertaken at the site;
- Any drainage or construction plans available for the site;
- Any Discharge Consents issued by the Environment Agency or the National Rivers Authority;
- Any trade effluent discharge consents issued by the local sewerage undertaker;
- Any records of any monitoring undertaken on any trade effluent discharge consent;
- Any water abstraction licenses held by the company;
- Any Authorisations granted by the Environment Agency/Her Majesties Inspectorate of Pollution for the operation/use of a Part A process/substance;
- Any Authorisations granted by the Local Environmental Health Department for the operation of a Part B process;
- Any records of any air emission monitoring undertaken at the site;
- Details of any solvent management system in place;
- Any Waste Management License held by the site;
- Any Registered Waste Carrier certificates held by the site;
- Any Special Waste Consignment Notes held on site;
- Any Duty of Care Transfer Noted held on site;
- Any Petroleum Licenses held on site;
- Any records relating to tank testing undertaken on the site;
- Any filling procedures held on site;
- Details of any emergency response plans in the event of any spillages on site;

- Any Consent issued under the Planning (Hazardous Substances) Act 1992;
- Details of any emergency plans etc. required if the site is registered under the Control of Industrial Major Accident Hazards Regulations;
- Any consent issued under the Notification of Installations Handling Hazardous Substances Regulations;
- Details relating to any requirements imposed on the company under the Dangerous Substances (Notification and Marking of Sites) Regulations;
- Information relating to whether the site is registered under the Radioactive Substances Act 1993;
- The sites COSHH manual and any associated Material Safety Data Sheets;
- Any asbestos survey reports that may have been carried out on site;
- Details of any noise monitoring that may have been undertaken on site;
- Any letters of complaint relating to the environmental effects of the site (e.g. complaints about noise, odour and dust from the site);
- Details of any fines, prosecutions or judgements made against the site for breaking any relevant environmental laws or regulations; and,
- Details of any future environmental expenditure and budgetary proposals or anticipated capital and revenue expenditure required in the next five years in order to comply with any environmental consents.

· Contaminated Land Regime ·

Contaminated Land (England) Regulations 1999, reg 2

2. (1) Contaminated land of the following descriptions is prescribed for the purposes of section 78C(8) as land required to be designated as a special site –

(a) land to which regulation 3 applies:

(b) land which is contaminated land by reason of waste acid tars in, on or under the land;

(c) land on which any of the following activities have been carried on at any time –

(i) the purification (including refining) of crude petroleum or of oil extracted from petroleum, shale or any other bituminous substance except coal, or

(ii) the manufacture or processing of explosives;

(d) land on which a prescribed process designated for central control has been or is being carried on under an authorisation where the process does not comprise solely things being done which are required by way of remediation;

(e) land within a nuclear site;

(f) land owned or occupied by or on behalf of –

(i) the Secretary of State for Defence;

(ii) the Defence Council;

(iii) an international headquarters or defence organisation; or

(iv) the service authority of a visiting force, being land used for naval, military or air force purposes;

(g) land on which the manufacture, production or disposal of –

(i) chemical weapons;

(ii) any biological agent or toxin which falls within section 1(1)(a) of the Biological Weapons Act 1974; or

(iii) any weapon, equipment or means of delivery which falls within section 1(1)(b) of that Act,

has been carried on at any time;

(h) land comprising premises which are or were designated by the Secretary of State by an order made under section 1(1) of the Atomic Weapons Establishment Act 1991;

(i) land to which section 30 of the Armed Forces Act 1996 applies; and

(j) land which –

(i) is adjoining or adjacent to land of a description specified in sub-paragraphs (b) to (i) above; and

(ii) is contaminated land by virtue of substances which appear to have escaped from land of such a description.

(2) For the purposes of paragraph (1)(b) above, "waste acid tars" are tars which –

(a) contain sulphuric acid;

(b) were produced as a result of the refining of benzole, used lubricants or petroleum; and

(c) are or were stored on land used as a retention basin for the disposal of such tars.

(3) In paragraph (1)(d) above, "authorisation" and "prescribed process" have the same meaning as in Part I of the Environmental Protection Act 1990 (integrated pollution control and air pollution control by local authorities) and the reference to designation for central control is a reference to designation under section 2(4) (which provides for processes to be designated for central or local control).

(4) In paragraph (1)(e) above, "nuclear site" means –

(a) any site in respect of which or part of which a nuclear site licence is for the time being in force; or

(b) any site in respect of which, or part of which, after the revocation or surrender of a nuclear site licence, the period of responsibility of the licensee has not come to an end;

and "nuclear site licence", "licensee" and "period of responsibility" have the meaning given by the Nuclear Installations Act 1965.

Table A – Categories of significant harm

Type of Receptor	Description of harm to that type of receptor that is to be regarded as significant harm
1 Human beings	Death, disease, serious injury, genetic mutation, birth defects or the impairment of reproductive functions. For these purposes, disease is to be taken to mean an unhealthy condition of the body or a part of it and can include, for example, cancer, liver dysfunction or extensive skin ailments. Mental dysfunction is included only insofar as it is attributable to the effects of a pollutant on the body of the person concerned. . . this description of significant harm is referred to as a "human health effect".
2 Any ecological system, or living organism forming part of such a system, within a location which is: • an area notified as an area of special scientific interest under section 28 of the Wildlife and Countryside Act 1981; • any land declared a national nature reserve under section 35 of that Act; • any area designated as a marine nature reserve under section 36 of that Act; • an Area of Special Protection for Birds, established under section 3 of that Act;	Harm which results in an irreversible adverse change, or in some other substantial adverse change, in the functioning of the ecological system within any substantial part of that location. In determining what constitutes such harm, the local authority should have regard to the advice of English Nature and to the requirements of the Conservation (Natural Habitats, etc) Regulations 1994. . . . this description of significant harm is referred to as an "ecological system effect".

Type of Receptor	Description of harm to that type of receptor that is to be regarded as significant harm
• any European Site within the meaning of regulation 10 of the Conservation (Natural Habitats etc) Regulations 1994 (ie Special Areas of Conservation and Special Protection Areas); • any habitat or site afforded policy protection under paragraph 13 of Planning Policy Guidance Note 9 (PPG9) on nature conservation (ie candidate Special Areas of Conservation, potential Special Protection Areas and listed Ramsar sites); or • any nature reserve established under section 21 of the National Parks and Access to the Countryside Act 1949	
3 Property in the form of: • crops, including timber; • produce grown domestically, or on allotments, for consumption; • livestock; • other owned or domesticated animals; • wild animals which are the subject of shooting or fishing rights.	For crops, a substantial diminution in yield or other substantial loss in their value resulting from death, disease or other physical damage. For domestic pets, death, serious disease or serious physical damage. For other property in this category, a substantial loss in its value resulting from death, disease or other serious physical damage. The local authority should regard a substantial loss in value as occurring

Type of Receptor	Description of harm to that type of receptor that is to be regarded as significant harm
	only when a substantial proportion of the animals or crops are dead or otherwise no longer fit for their intended purpose. Food should be regarded as being no longer fit for purpose when it fails to comply with the provisions of the Food Safety Act 1990. Where a diminution in yield or loss in value is caused by a pollutant linkage, a 20% diminution or loss should be regarded as a benchmark for what constitutes a substantial dimunition or loss. . . . this description of significant harm is referred to as an "animal or crop effect".
4 Property in the form of buildings. For this purpose, "building" has the meaning given in section 336(1) of the Town and Country Planning Act 1990 (ie it includes "any structure or erection, and any part of a building . . . but does not include plant or machinery comprised in a building").	Structural failure, substantial damage or substantial interference with any right of occupation. For this purpose, the local authority should regard substantial damage or substantial interference as occurring when any part of the building ceases to be capable of being used for the purpose for which it is or was intended. Additionally, in the case of a scheduled Ancient Monument, substantial damage should be regarded as occurring when the damage significantly impairs the historic, architectural, traditional, artistic or archaeological interest by reason of which the monument was scheduled. . . . this description of significant harm is referred to as a "building effect".

Guidance on Assessment · and Redevelopment of · Contaminated Land

Table 3 Tentative "trigger concentrations" for selected inorganic contaminants.

Conditions

1. All values are for concentrations determined on "spot" samples based on adequate site investigation carried out prior to development. They do not apply to analysis of averaged, bulked or composited samples, nor to sites which have already been developed. All proposed values are tentative.

2. The lower values in Group A are similar to the limits for metal content of sewage sludge applied to agricultural land. The values in Group B are those above which phytotoxicity is possible.

3. If all samples are below the threshold concentrations then the site may be regarded as uncontaminated as far as the hazards from these contaminants are concerned and development may proceed. Above these concentrations, remedial action may be needed especially if the contamination is still continuing. Above the action concentration, remedial action will be required or the form of development changed.

CONTAMINANTS	PLANNED USES	TRIGGER concentrations (mg/kg air-dried soil)	
		Threshold	Action
Group A: Contaminants which may pose hazards to health			
Arsenic	Domestic gardens, allotments.	10	•
	Parks, playing fields, Open space.	40	•
Cadmium	Domestic gardens, allotments.	3	•
	Parks, playing fields, Open space.	15	•
Chromium (hexavalent) (1)	Domestic gardens, allotments. Parks, playing fields, Open space.	25	•
Chromium (Total)	Domestic gardens, allotments.	600	•
	Parks, playing fields, Open space.	1000	•
Lead	Domestic gardens, allotments.	500	•
	Parks, playing fields, Open space.	2000	•
Mercury	Domestic gardens, allotments.	1	•
	Parks, playing fields, Open space.	20	•
Selenium	Domestic gardens, allotments.	3	•
	Parks, playing fields, Open space.	6	•
Group B: Contaminants which are phytotoxic but not normally hazardous to health.			
Boron (water soluble) (3)	Any uses where plants are to be grown (2,6)	3	•
Copper (4,5)	Any uses where plants are to be grown (2,6)	130	•
Nickel (2,6)	Any uses where plants are to be grown (2,6)	70	•
Zinc (4,5)	Any uses where plants are to be grown (2,6)	300	•

Notes:

(1): Action concentrations will be specified in the next edition of ICRCL 59/83

(2): Soluble hexavalent chromium extracted by 0.1% HCI at 37°C; solution adjusted to pH 1.0 if alkaline substances present.

(3): The soil pH value is assumed to be about 6.5 and should be maintained at this value. If the pH falls, the toxic effects and the uptake of these elements will be increased.

(4): Determined by standard ADAS method (soluble in hot water).

(5): Total concentration (extractable by HNO3/HCLO4).

(6): The phytotoxic effects of copper, nickel and zinc may be additive. The trigger values given here are those applicable to the 'worst case':phytotoxic effects may occur at these concentrations in acid, sandy soils. In neutral or alkaline soils phytotoxic effects are unlikely at these concentrations.

(7): Grass is more resistant to phytotoxic effects than are most other plants and its growth may not be adversely affected at these concentrations.

Table 4 Tentative "trigger concentrations" for contaminants associated with former coal carbonisation

Conditions

1. All values are for concentrations determined on "spot" samples based on adequate site investigation carried out prior to development. They do not apply to analysis of averaged, bulked or composited samples, nor to sites which have already been developed. All proposed values are tentative.

2. Many of these values are preliminary and will require regular updating. They should not be applied without reference to the current edition of the report "Problems Arising from the Redevelopment of Gas Works and Similar Sites".

3. If all samples are below the threshold concentrations then the site may be regrded as uncontaminated as far as the hazards from these contaminants are concerned and development may proceed. Above these concentrations, remedial action may be needed, especially if the contamination is still continuing. Above the action concentration, remedial action will be required or the form of development changed.

CONTAMINANTS	PLANNED USES	TRIGGER CONCENTRATIONS (mg/kg air-dried soil)	
		Threshold	Action
Polyaromatic hydrocarbons (1,2)	Domestic gardens, allotments, play areas.	50	500
	Landscaped areas, buildings, hard cover.	1000	10000
Phenols	Domestic gardens, allotments.	5	200
	Landscaped areas, buildings, hard cover.	5	1000
Free cyanide	Domestic gardens, allotments, landscaped areas.	25	500
	Buildings, hard cover.	100	500
Complex cyanides	Domestic gardens, allotments.	250	1000
	Landscaped areas.	250	5000
	Buildings, hard cover.	250	NL
Thiocyanate (2)	All proposed uses.	50	NL
Sulphate	Domestic gardens, allotments, landscaped areas.Buildings (3). Hard cover.	2000	10000
		2000(3)	50000(3)
		2000	NL
Sulphide	All proposed uses.	250	1000
Sulphur	All proposed uses.	5000	20000
Acidity (pH less than)	Domestic gardens, allotments, landscaped areas.	pH5	pH3
	Buildings, hard cover.	NL	NL

Notes:

NL: No limit set as the contaminant does not pose a particular hazard for this use.

(1): Used here as a marker for coal tar, for analytical reasons. See "Problems Arising from the Redevelopment of Gas Works and Similar Sites" Annex A1. (1).

(2): See "Problems Arising from the Redevelopment of Gas Works and Similar Sites" for details of analytical methods. (1).

(3): See also BRE digest 250: Concrete in sulphate-bearing soils and ground water.

Source: ICRCL guidance note 59/83 second edition July 1987. The reader is advised to consult this document.

Dutch intervention values for soil and water contaminants (Van den Berg et al, 1993)

Substance		Concentration in soil: mg/kg dry weight		Concentration in groundwater: µg/l	
		T	I	T	I
Metals	Chromium	100	380	1	30
	Cobalt	20	240	20	100
	Nickel	35	210	15	75
	Copper	36	190	15	75
	Zinc	140	720	65	800
	Arsenic	29	55	10	60
	Molybdenum	10	200	5	300
	Cadmium	0.8	12	0.4	6
	Barium	200	625	50	625
	Mercury	0.3	10	0.05	0.3
	Lead	85	530	15	75
Inorganic pollutants	Cyanides (free)	1	20	5	1500
	Cyanides – complex (pH<5)[1]	5	650	10	1500
	Cyanides – complex (pH≥5)[1]	5	50	10	1500
	Thiocynates	–	20	–	1500
Aromatic compounds	Benzene	0.05	1	0.2	30
	Ethyl benzene	0.05	50	–	150
	Phenol	0.05	40	–	2000
	Cresol	–	5	–	200
	Toluene	0.05	130	0.2	1000
	Xylene	0.05	25	0.5	70
	Catechol	–	20	–	1250
	Resorcinol	–	10	–	600
	Hydrochinon	–	10	–	800
Polycyclic aromatic compounds (PAHs)	Total PAH (sum of 10)	1	40[2,11]	–	–
	Naphthalene	0.015	–	0.1	70
	Anthracene	0.05	–	0.02	5
	Phenanthrene	0.045	–	0.02	5
	Fluoranthene	0.015	–	0.005	1
	Benzo(a)anthracene	0.02	–	0.002	0.5
	Chrysene	0.02	–	0.002	0.05
	Benzo(a)pyrene	0.02	–	0.001	0.05
	Benzo(ghi)perylene	0.02	–	0.0002	0.05
	Benzo(k)fluoranthene	0.025	–	0.001	0.05
	Indeno(1,2,3,cd)pyrene	0.025	–	0.0004	0.05

Substance		Concentration in soil: mg/kg dry weight		Concentration in groundwater: µg/l	
		T	I	T	I
Chlorinated organic compounds	1,2-Dichloroethane	–	4	(d)	400
	Dichloromethane	–	20	(d)	1000
	Tetrachloromethane	0.001	1	(d)	10
	Tetrachloroethane	0.01	4	(d)	40
	Trichloromethane	0.001	10	(d)	400
	Trichloroethene	0.001	60	(d)	500
	Vinyl chloride	–	0.1	–	0.7
	Chlorobenzene (sum)	–	30[3,11]	–	–
	Monochlorobenzene	(d)–		0.01	180
	Dichlorobenzene	0.01	–	0.01	50
	Trichlorobenzene	0.01	–	0.01	10
	Tetrachlorobenzene	0.01	–	0.01	2.5
	Pentachlorobenzene	0.025	–	0.01	1
	Hexachlorobenzene	0.025	–	0.01	0.5
	Chlorophenols (sum)	–	10[4,11]	–	–
	Monochlorophenol	0.0025	–	0.25	100
	Dichlorophenol	0.003	–	0.08	30
	Trichlorophenol	0.001	–	0.025	10
	Tetrachlorophenol	0.001	–	0.025	10
	Pentachlorophenol	0.002	5	0.013	
	Chloronaphthalene	–	10	–	6
	Polychlorobipheyls (sum of 7)	0.02	15	(d)	0.015
Pesticides	DDT/DDD/DDE	–	4[6]	–	0.016
	Drins (sum of 3)	–	4[7]	–	0.017
	HCH – compounds	–	2[8]	–	18
	Carbaryl	–	5	–	0.1
	Carbofuran	–	2	–	0.1
	Manab	–	35	–	0.1
	Atrazine	0.00005	6	0.0075	150
Other pollutants	Cyclohexanes	–	270	–	15000
	Phthalates (sum)	0.1	60[9]	0.5 –	5[9]
	Mineral Oil	50	5000[10]	50	600[10]
	Pyridine	0.1	1	0.5	3
	Styrene	0.1	100	0.5	300
	Tetrahydrofuran	–	0.4	–	1
	Tetrahydrothiophene	–	90	–	30

Greater London Council Guidelines for Contaminated Soils

Suggested range of values (mg/kg on air dried soils, except for pH)

Parameter	Typical values for uncontaminated land	Slight Contamination	Contaminated	Heavy Contamination	Unusually Heavy Contamination
pH (acid)	6–7	5–6	4–5	2–4	<2
pH (alkali)	7–6	8–9	9–10	10–12	12
Antimony	0–30	30–50	50–100	100–500	500
Arsenic	0–30	30–50	50–100	100–500	500
Cadmium	0–1	1–3	3–10	10–50	50
Chromium	0–100	100–200	200–500	500–2500	2500
Copper – available	0–100	100–200	200–500	500–2500	2500
Lead	0–500	500–1000	1000–2000	2000–1.0%	1.0%
Lead – available	0–200	200–500	500–1000	1000–5000	5000
Mercury	0–1	1–3	3–10	10–50	50
Nickel – available	0–20	20–50	50–200	200–1000	1000
Zinc – available	0–250	250–500	500–1000	1000–5000	5000
Zinc – equivalent	0–250	250–500	500–2000	2000–1%	1.0%
Boron – available	0–2	2–5	5–50	50–250	250
Selenium	0–1	1–3	3–10	10–50	50
Barium	0–500	500–1000	1000–2000	2000–1.0%	1.0%
Berylium	0–5	5–10	10–20	20–50	50
Manganese	0–500	500–1000	1000–2000	2000–1.0%	1.0%
Vanadium	0–100	100–200	200–500	500–2500	2500
Magnesium	0–500	500–1000	1000–2000	2000–1.0%	1.0%
Sulphate	0–2000	2000–5000	5000–1.0%	1.0–3.0%	3.0%
Sulphur – free	0–100	100–500	500–1000	1000–5000	5000
Sulphide	0–10	10–20	20–100	100–500	500
Cyanide – free	0–1	1–5	5–50	50–100	100
Cyanide – total	0–5	5–25	25–250	250–500	500
Ferricyanide	0–100	100–500	500–1000	1000–5000	5000
Thiocyanate	0–10	10–50	50–100	100–500	2500
Coal tar	0–500	500–1000	1000–2000	2000–1.0%	1.0%
Phenol	0–1	2–5	5–50	50–250	250
Toluene Extractable Matter	0–5000	5000–1.0%	1.0–5.0%	5.0–25.0%	25.0%
Cyclohexane Extractable Matter	0–2000	2000–5000	5000–2.0%	2.0–10.0%	10.0%

Source: Site investigation and materials problems, proc. Conference on Reclamation of Contaminated Land, Eastbourne, October 1979 (Society of Chemical Industry, London 1980).

· Index ·

Abatement notice,
    service of, 34
Abstraction. *See* **Licensed water abstraction, Water abstraction**
Acids,
    effect of, 137
Acquisition audits. *See also* **Pre-acquisition audits**
    operation of, 3, 10
Action programmes,
    progress, made through, 19, 20
Air pollution,
    action taken against, 20
    guidance, issued on, 192, 197
Air Pollution Control,
    local authority, position of, 22, 24
Air quality,
    legislation affecting, 41, 56
Air sparging,
    remedial action, involving, 148
Alkalis,
    effect of, 137
Anti-pollution works notice,
    service of, 37
Aquifers,
    definition of, 143
    importance of, 143
    location of, 143

Asbestos,
    definition of, 186
    risks, associated with, 139
Assessment procedures,
    operation of, 2, 3
Audit objectives,
    international standards, achieving, 153
Audit process, 58 *et seq*
    audit evidence,
        evaluation of, 66
        gathering of, 64, 65

audit team,
        resources, assessment of, 62
        selection of, 60
    communication methods,
        establishing, 65, 66
    exit meeting, conduct of, 66, 67
    facility information, required, 61
    health & safety considerations, 62
    hot work, permits required for, 62
    internal controls, appreciation of, 63, 64
    on-site activities, 61 *et seq*
        initial tour, carrying out, 61–63
        opening meeting, 61, 62
    management systems, appreciation of, 63
    post-audit activities, 67, 68
    pre-audit activities,
        identifying audit scope, 59
        informing management, 59
        planning routines, 59, 60
        selecting facilities, 58
    risk assessment, importance of, 63
    site information, required, 61
    site restrictions, considered, 62
    verification, procedures for, 65
    working papers, required, 60
Audit programme,
    function of, 2
    internal resources, consideration of, 5
    structure, considerations as to, 5
Audit protocols, 104 *et seq*
    checklists, use of, 105, 106,
        basic checklist, 106, 107
        detailed checklist, 106, 108
        topical checklist, 106, 108
    form of, 105
    functions of, 105
    preparation of, 105
Audit work programme, 104 *et seq*
Auditing. *See also* **Compliance auditing**
    definition of, 1, 2, 4

Auditing techniques,
    corporate activity, involving, 1
    development of, described, 1, 6, 7
    documentation, attention paid to, 3
    function of, 1
    nature of, 3, 4
    requirements for, 3, 4
    verification, by means of, 3
Auditing types,
    categories described, 4
    choice of, decision as to, 6
    classification of, 5
Auditors,
    international standards, relating to,
        154
Authorisation,
    toxic substances, involving, 36

BAT,
    "available", meaning of, 28
    "best", meaning of, 28
    costs, associated with, 28
    existing processes, dealing with, 28,
        29
    new processes, affecting, 28
    practical implications, associated
        with, 27
    standards, establishment of, 28, 29
BATNEEC,
    continuing harm, where, 28
    implementation of, 23, 24, 27
    requirements, relating to, 27
    standards, establishment of, 29
Bibliography, 165 *et seq*
Bioremediation techniques, 148, 150
Biotechnology and Biotechnological
    Sciences Research Council, 50
BPEO,
    achievement of, 23, 29
Business in the environment, 53

Chemical Industries Association,
    contribution made by, 8, 53
Chemical industry,
    pollution control, relating to, 193,
        196
Coal tar,
    effects of, 138
Commercial property,
    audits involving, 5
Communication skills,
    importance attached to, 65, 66

Company management,
    audit process, identification with, 63,
        112
    management standards, applied, 116
    resources, company use of, 5, 64
    training standards, applied, 116
Company performance,
    standards, measured against, 4
Company policy,
    assessment of, 7, 116
Compliance,
    corporate responsibilities, towards, 1
    policy guidance, concerning, 7, 8
    statutory requirements, relating to, 6,
        7
Compliance auditing,
    definition of, 2
    nature of, 7
    practice, relating to, 7, 8
        auditor fatigue, consideration
            given to, 9
        duplication, avoidance of, 8
        logistics, importance of, 9
        management resistance, dealing
            with, 9
        multi-functional approach, 8, 9
        specialist knowledge, applied, 8
        time, effective use of, 8, 9
    structure, considerations as to, 7
Compliance status,
    company responsibilities, concerning,
        1
    information provided, on, 128
Confederation of British Industry (CBI),
    contribution made by, 8
Construction debris,
    definition of, 187
Contaminated land,
    acids, effect of, 137
    alkalis, effect of, 137
    cyanides, use of, 138
    definition of, 42
    deleterious materials, effect of, 135,
        136
    development of, 215, 216
    disclosure provisions, relating to, 132
    gas transmissions, leakage from, 134
    ground conditions, state of, 135
    hazardous materials, effects of, 136
    health risks, associated with, 42, 43
    industries, associated with, 140–142
    landfill sites, redevelopment of, 133

legislative measures, controlling, 41, 42, 132, 210
liability,
  apportionment of, 47
  "caused and knowingly permitted", 46
  establishing liability, means of, 45, 46
  exemptions from, 46
  extent of, for, 43
  liability groups, importance of, 46, 47
licences, provision for, 31
local authorities, involvement of, 43
metal compounds, use of, 137
notification procedures, 46
organic compounds, use of, 138
organic contaminants, decomposition of, 134
planning conditions, involving, 43
pollutant linkage, consideration given to, 44, 45
provisions affecting, 26
remediation,
  action involving, 46, 47, 56, 132
  costs, associated with, 43, 45
  notice, service of, 47, 48
responsibilities involving, 45
risks, associated with,
  explosive gasses, 133
  flammable gasses, 133
sources, contamination of, 140
tests to establish, 43–45
toxic materials, effect of, 136
valuation issues, involving, 43
Contaminative use,
  categories of, 189, 190, 201 *et seq*
    ceramics, making of, 202
    chemicals, use of, 203
    energy industry, relating to, 201
    engineering processes, 203
    extractive industry, relating to, 201
    food processing, relating to, 203
    glass making processes, 202
    infrastructure, relating to, 204, 205
    manufacturing processes, 203
    metals, production of, 202
    non-metals, production of, 202
    rubber industry, relating to, 204
    textile processes, involving, 204
    timber products, relating to, 204
    waste disposal, relating to, 205

Contingency planning,
  consideration given to, 116
Corrective action,
  reference made to, 129
Costs,
  considerations, as to, 27, 28
Countryside Commission, 50
Countryside Council of Wales, 50
Cyanides,
  use of, 138

Deleterious materials,
  effect of, 135, 136
Design operation,
  attention paid to, 120, 121
Developing countries,
  interests of, considered, 19
Discharge,
  levels of, assessment of, 121, 123
Discharge consents,
  information, relating to, 97
Disposal,
  problems, associated with, 118, 119
Distribution systems,
  attention paid to, 121, 122
Divestment,
  audits, in relation to, 11
Documentation,
  importance, attached to use, 3, 104
Drains, attention paid to, 117
Due diligence. *See also* **Due diligence inquiry**
  practice relating to, 3, 10, 11, 41
Due diligence inquiry,
  air pollution, details of, 185
  breach of regulations, details of 183, 184
  consents pertaining, details of, 183
  contaminated material, existence of, 182, 184
  information required,
    environmental liability, 179
    hazardous substances, concerning, 179, 181
    insurance policies, availability of, 186
    land use, regarding, 178, 179
    location of land, 180
    register, whether held on, 182
  litigation pending, whether, 182
  noise levels, information on, 186
  pollution control, details concerning, 185

previous investigations, whether held, 180

radioactive substances, details of, 185

waste management, information concerning, 185

water abstractions, details of, 181

Duty of care,
  principles involving,
    waste management, 31, 32, 185

Ecological balance,
  importance attached to, 19

Effects. *See* **Environmental effects**

Effluent,
  discharge of, improvements sought, 123

Electrocoagulation techniques, 148, 149

Emissions,
  levels of, assessment of, 121, 123

Employers,
  health and safety, responsibilities of, 40

Energy use,
  assessment of, 122
  costs, associated with, 6

English Nature, 50, 51

Environment Agency,
  contact details, 170 *et seq*
  contaminated land, action determined by, 44, 45
  controlled waters, responsibility for, 35
  creation of, 41, 55
  enforcement powers of, 36, 37
  fisheries management, involving, 56
  flood defences, involving, 56
  information, sought from, 102
  objectives, set by, 23, 26
  registers, maintained by, 56
  responsibilities of, 55
  sustainable development, encouraged by, 55, 56
  work of, 21, 22, 27, 56

Environment Council, 53

Environmental action programmes,
  development of, 18
  objectives, set for, 18

Environmental audit. *See also* **Auditing**
  meaning of, 187

Environmental Crime Unit,
  work of, 1

Environmental Data Association (EDA),
  role of, 92

Environmental data providers,
  information, collected from, 81 *et seq*
  quality of information, acquired from, 93
  regulations affecting, 92

Environmental effects,
  assessment of,
    danger areas, 117
    problem areas, 117
  identification of, 116, 118

Environmental impact,
  environmental impact statement,
    industrial processes, for, 30

Environmental Industries Council, 53

Environmental liability,
  due diligence inquiry, assessment of, 179

Environmental Management System (EMS),
  international standards, relating to, 153

Environmental performance,
  monitoring of, 126, 129

Environmental protection,
  legislation affecting, development of, 15 *et seq*

Environmental Protection Agency,
  policy, origins derived from, 7, 8

European Environment Agency,
  role of, 57

Fisheries management,
  responsibility for, 56

Flood defence,
  responsibility for, 56

Friends of the Earth, 54

Fuel production processes,
  pollution control, issues concerning, 192, 195

Gas transmissions,
  leakage from, 134

Geological data,
  use of, 94

Glossary of terms, 156 *et seq*

Green Alliance, 54

Green audit,
  nature of, 13

Greenpeace, 54

Ground conditions,
  state of, 135

Groundwater, 142 *et seq*,
  abstraction of, 147, 148

definition of, 36
flow of, 142, 144
hydrological cycle, effect of, 142
legislative measures, concerning, 35,
    36
remedial, action involving, 147, 148,
    150
sources of, 142
Guidance notes,
    importance attached to, 21, 24, 26,
        28, 30

Hazardous materials,
    effects of, 136
Hazardous substances,
    code of practice, concerning, 39
    consents, provisions covering, 37
    definition of, 187
    duties, in relation to, 38
    health, damaging to, 38, 39
    legislation covering, 37 *et seq*
    notification procedures, for, 37, 38
    quantities, relevant levels, 38
Health and safety,
    audit process, consideration given to,
        62, 116
    responsibilities involving, 2, 8
Health and Safety Executive,
    work of, 51
Health risk,
    assessment of, 40, 42, 43
Historical maps,
    use of, 69, 70, 80
HM Inspectorate of Pollution (HMIP),
    work of, 22, 34
Hydrogeology,
    use of, 96
Hydrology,
    use of, 96

**Industry Council for Packaging and the
    Environment,** 54
Information,
    audit process, handling of, 61, 69, 70,
        80, 100
    pre-survey questionnaires, use of, 98,
        99
    protocols, use of, 104, 108
    register, held on, 182
    sources, access to, 81 *et seq*, 102
Information request forms,
    use of, 100

Institute for European Environmental
    Policy, 54
Insurance,
    auditing connected with, 11, 12
    policies, whether available, 186
Integrated Pollution Control (IPC),
    authorisation, relating to, 25
    guidance relating to, 24, 28, 30
    legislation affecting, 21, 22, 25
    objectives, relating to, 22, 23
Internal control questionnaire,
    use of, 110, 111
International Chamber of Commerce
    (ICC),
    contribution made by, 3, 7
International standards,
    adherence to, 130, 131, 151, 154
    EN ISO 14010: 1996, 151, 152
    EN ISO 14011: 1996, 153
    EN ISO 14012: 1996, 154
IPPC system,
    cost, considerations as to, 27
    effect of, 25
    guidance notes, issue of, 26
    installation, definition of, 26
    introduction of, 25

Joint Nature Conservation Committee,
    50

Land use,
    due diligence inquiry, details of, 178,
        179
Landfill sites,
    redevelopment of, provisions
        affecting, 133
Legislation,
    air quality, relating to, 41, 56
    contaminated land, involving, 132,
        210
    development of, 15, 21
    devolution, effect of, 25
    European influences on, 15, 16, 18
    guidance notes, issue of, 21, 24, 26,
        30
    national parks, involving, 41
    on-site storage, for, 120
    packaging materials, affected by, 119
    pollution control, covering, 30
    waste management, affecting, 124
    water, measures involving, 34–36
Liability. *See also* **Product liability**
    divestment audits, 11

Liability audits,
  practice relating to, 3, 10–12
  pre-acquisition audits, 3, 10
Licence,
  meaning of, 189
Licensed water abstraction,
  information, relating to, 96, 102
Life cycle assessment,
  elemental flows, identification of, 14
  practice relating to, 13, 14
Local authorities,
  abatement notice, served by, 34
  air quality reviews, carried out by, 41
  contaminated land, responsibility for,
    43
Local Authorities Air Pollution Control
  (LAAPC), 22, 24
  role of, examined, 21, 25

Maastricht Treaty,
  effect of, 17, 18
Management responsibility,
  auditing,
    identification with, 112
    resistance towards, 9
Manhole covers,
  attention paid to, 117
Manufactured goods,
  auditing requirements, affecting, 13
Manufacturing process,
  environmental impact, associated
    with, 14
Maps. *See* **Historical maps**
Marine Conservation Society, 54
Metal compounds,
  effect of, 137
Metals processing,
  pollution control, relating to, 193,
    195
Methane,
  risks, associated with, 133
Methodology,
  comprehensive approach, need for,
    104
  systematic approach, need for, 104
Mineral industry,
  pollution control, involving, 193, 195

National Parks,
  legislation affecting, 41
National programmes,
  co-ordination of, 19

National Rivers Authority (NRA),
  creation of, 34
  responsibilities of, 34
National Society for Clean Air &
  Environmental Protection, 54, 55
National Trust, 55
Natural Environment Research Council,
  51
Natural resources,
  exploitation of, 19
  sustainable management of, 21
Neighbouring property,
  definition of, 189, 190
Noise pollution,
  action taken against, 20
Non-Governmental Organisations
  (NGOs),
  role of, 53 *et seq*
Noxious gases,
  definition of, 188
Noxious odours,
  attention paid to, 117
Nuisance,
  prevention at source, 18, 19

Official publications, 166, 167
On-site storage,
  fugitive emissions, existence of, 120
  legal requirements, as to, 120
  responsibility for, 120
  risks, associated with, 120
  suitability, standards for, 120
Open-ended questionnaires,
  use of, 109, 110
Operational factors,
  dealing with,
    contingency planning, 116
    health & safety record, 116
    management standards, 116
    packaging materials, 119
    pollution abatement equipment,
      116, 117
    training standards, 116
  disposal, problems associated with,
    118, 119
  environmental effects, assessment of,
    118
  product design, attention paid to,
    118, 119
Operations,
  management understanding, of, 112
Oral reports,
  use made of, 127, 128

Organic compounds,
  effect of, 138
Organic contaminants,
  decomposition of, 134

Packaging materials,
  company use, assessment of, 119
  legislation covering, 119
  recycled material, proportion of, 119
Packaging waste,
  responsibility for, 13, 14
PCBs,
  effects of, 138
Performance assessment,
  need for, 129
Petroleum hydrocarbons,
  risks, associated with, 134, 135
Phenols,
  effects of, 138
Planning records,
  use of, 70 *et seq*
Pollutant pathways,
  boreholes, requirements for, 145
  pollution,
    distributed sources, 144
    point sources, 144
Polluter,
  polluter pays, principle of, 19, 20
Pollution. *See also* **Pollutant pathways,
  Pollution control**
  liquids causing, 145
  pollution abatement equipment,
    condition of, 116, 117
  prevention at source, 18, 19
  protection against, 19
  soluble characteristics, 145
  sources of, 144, 145
Pollution control. *See also* **IPPC system**
  industrial processes, subject to, 192,
    193, 195, 196
  legislation affecting, 22, 25, 30
  phase-in dates, details of, 199, 200
  waste disposal, subject to, 194, 197
Pre-acquisition inquiry. *See also* **Due
  diligence inquiry**
  information, required for, 101, 178 *et
    seq*
Prescribed process,
  legislation affecting, 30
  meaning of, 190
Pre-survey questionnaires,
  use of, 98, 99, 206, 207

Process diagrams,
  interpretation of, 124
  use of, 124, 125
Process operation,
  attention paid to, 120
  discharge involved, levels of, 121, 123
  emissions involved, level of, 121, 123
Product design,
  attention paid to, 118, 119
  legislation affecting, 119
  processes, involved with, 118, 119
  recycling, potential for, 119
Product liability,
  life cycle analysis, 13, 14
  suppliers' undertakings, as to, 13
Property transactions,
  audits, conducted for, 4
  due diligence, practice involving, 10,
    11
Protection. *See* **Environmental protection**
Protocols. *See also* **Audit protocols,
  Questionnaire protocols**
  use of, 104 *et seq*
Public awareness,
  importance of, 19

Questionnaire protocols,
  use of, 108, 109
Questionnaires. *See* **Pre-survey
  questionnaires, Questionnaire
  protocols**

Radioactive waste,
  information on, requirements as to,
    185
  legislative measures affecting, 37, 56
Raw materials,
  storage of, 120
  use of, levels involved, 123
Regulatory authorities, 48 *et seq*
  central government departments,
    DETR, work of, 48, 49
    DTI, work of, 49
    MAAF, work of, 49
  local government, 52 *et seq*
    county councils, work of, 52
    district council, work of, 52
    unitary authorities, work of, 52,
      53
  national bodies, work of, 49, 50
Reporting mechanisms,
  audit team, information on, 130

oral reports, use of, 127, 128
performance assessment, need for,
    129
practice relating to, 127 *et seq*
priorities, allocation of, 130
recommendations given, 129
reporting criteria, requirements as to,
    131
written reports, use of, 128, 129
Resources,
    company resources, assessment of, 5,
    7
Retail quantities,
    meaning of, 190
Review procedures,
    operation of, 2
Rio Conference (1992),
    sustainable development, envisaged
        by, 21
Risk assessment,
    audit process, as part of, 63
    requirements, relating to, 12
Royal Commission on Environmental
    Pollution, 51
Rural areas,
    action taken, with regard to, 20

Scored questionnaires,
    use of, 110
Scotland,
    hazardous substances, measures
        affecting, 37, 38
    pollution control, legislation
        affecting, 22, 25
Scottish Environmental Protection
    Agency,
    contact details, 174
    work of, 41, 56, 57
Scottish National Heritage,
    work of, 51
Sewers,
    substances discharged, measures
        controlling, 35
Significant harm,
    categories of, 212–214
Single issue audits,
    practice relating to, 9 *et seq*
Site. *See also* **On-site storage, Site
    condition report, Site management,
    Site remediation**
    abstract licensing, relating to, 102
    description of, need for, 129
    environmental health issues, 103

groundwater protection, matters
    concerning, 102
pollution control, issues involving,
    102
pre-survey questionnaires, use of, 98,
    99
site history, summary of, 79
site information,
    collecting information, on, 97, 98,
        100, 101, 104
    request forms, use of, 100
    trade effluent, relating to, 103
    uses made of, determining, 69
    waste regulation, 102
Site condition report,
    requirements for, 26
Site management,
    attention paid to, 122
Site remediation, 146 *et seq*
    groundwater, involving, 147, 148
    remedial action,
        cleaning up, 147
        covering up, 146
        criteria for, 146
        removal, 146
    soil, involving, 149
Soil,
    excavation of, 149
    guidelines, concerning, 220
    site remediation, involving, 149
Statutory nuisance,
    provisions, relating to, 33, 34
Storage. *See* **On-site storage**
Storage tanks,
    attention paid to, 117
Suppliers,
    performance of, monitoring, 126
    undertakings, given by, 13
Sustainable development,
    encouragement of, 55, 56
Sustainable management,
    natural resources, protection of, 21

Technical knowledge,
    application of, 19
Technical planning,
    environmental impact, assessed
        during, 18
Toxic materials,
    effects of, 136
Toxic substances,
    discharge of,
        authorisation required, 36

measures controlling, 35, 36
Trade effluent discharges,
    information, relating to, 97, 103
Transport,
    company practices, assessment of,
        121, 122
Treaty of Rome,
    influence of, 16
Types of audit. *See* **Auditing types**

UN Environmental Programme (UNEP),
    creation of, 16

Waste control. *See also* **Packaging waste**
Waste disposal,
    pollution control, exercised over, 194,
        197
    product description, relating to, 32
    statutory provisions, affecting, 12, 13
Waste generation,
    control, issues involving, 12
    costs, associated with, 6
Waste management,
    code of practice, involving, 32
    controls affecting, 124
    duty of care, relating to, 31, 32, 124
    legislation affecting, 30, 31, 124
    licence, required for, 31
    national strategy, for, 41, 42, 56
    producer responsibility, extent of, 31,
        32
    transfer of waste, 32, 56
waste,
    definition of, 32, 33
    description of waste, requirement
        for, 32
    disposal of, 32, 33
    household waste, responsibility
        for, 31
Water. *See also* **Groundwater, Water
    abstraction, Water pollution, Water
    use**
    controlled waters, provisions
        affecting, 35, 36
    legislative measures involving, 34 *et
        seq*
Water abstraction,
    due diligence inquiry, details of, 181
    licences involving, 96, 102
Water pollution,
    action taken against, 20
Working papers,
    action lists, use of, 115
    auditing process, involving, 111–113
    content of, 112, 113
    format for, 113, 114
    retention of, 114
    review of, need for, 113
    use of,
        on-site activity, 112
        pre-audit activity, 112
Written reports,
    use of, 128, 129